T0222288

Springer Undergraduate Mathematics Series

More information about this series at http://www.springer.com/series/3423

Alexander Isaev

Twenty-One Lectures on Complex Analysis

A First Course

 Springer

Alexander Isaev
Mathematical Sciences Institute
Australian National University
Acton, Aust Capital Terr, Australia

ISSN 1615-2085   ISSN 2197-4144   (electronic)
Springer Undergraduate Mathematics Series
ISBN 978-3-319-68169-6   ISBN 978-3-319-68170-2   (eBook)
DOI 10.1007/978-3-319-68170-2

Library of Congress Control Number: 2017958438

Mathematics Subject Classification (2010): 97Ixx, 97I80

Printed on acid-free paper

This Springer imprint is published by Springer Nature
The registered company is Springer International Publishing AG
The registered company address is: Gewerbestrasse 11, 6330 Cham, Switzerland

To Esya, who has always pulled me out of trouble

Preface

This book has grown out of an undergraduate course that I have taught at the Australian National University (ANU) for over 20 years. The course is one-semester long, which means that it runs for a total of 12 weeks. In each week I teach two lectures, where a lecture is defined as a 100-minute class with a 5-minute break in the middle. This lecture format is common in Europe but can be implemented almost everywhere; for example, at the ANU—where a normal teaching period is 50 minutes—I simply reserve two consecutive periods for every lecture. A 12-week semester can in principle accommodate 24 lectures of this kind but the course material only occupies 21, with the remaining time spent on discussing assignments, review, etc.

While I was transforming my lecture notes into a book, I decided to keep the splitting of the course material into 21 lectures. This approach is unusual as most authors would organise the content into chapters, with each chapter accommodating a particular topic. However, I find that dividing the material according to the way it is presented at the lectures has several advantages compared to the traditional topic-based book composition. Indeed, first of all, the lecture-based organisation ensures that the content is partitioned into (approximately) equal pieces, so that none of them stands out and looks intimidating to the students at least as far as the length is concerned. This issue becomes particularly important for those who wish to use the book for self-study and would like to keep a close eye on their overall progress. Secondly, the lecture-based format guarantees that the students get more training for the more advanced topics, which are spread over several lectures. Indeed, each lecture has its own unique set of exercises, and the students are strongly encouraged to do at least some of them before moving on. It is then automatic that the harder the topic, the more exercises one is expected to do to go through it. It should also be mentioned that some of the exercises for each lecture serve as a preparation for the following one. Thirdly, the lecture-based split-up gives clear teaching guidelines to the instructor, at the same time allowing for the possibility of re-arranging the material according to their own taste.

As the book covers a one-semester course, it is shorter than most complex analysis texts (approximately 200 pages). It is well-known that many students are intimidated by long large books, so this shorter one—which contains exactly the material that needs to be learned in a one-semester course—is expected to have a broader appeal. Another feature of the book that the students may like is a reader-friendly conversational style of writing. For instance, there are plenty of fully worked-out examples and textual explanations of formal statements, with plain words systematically used in the formulations of theorems, propositions, etc. Furthermore, the reader is invited to participate in the exposition by filling in various details of formal arguments as indicated by parenthesised expressions, e.g., (check!), (explain!), (provide details!). The proofs of some of the statements are left as homework. In fact, doing weekly homework is strongly encouraged with plenty of exercises to choose from at the end of each lecture. Note that the exercises have a varying degree of difficulty to accommodate different cohorts of students and range from routine questions to rather hard problems.

Although the choice of topics covered in the book may appear to be standard, this is more than just a book on complex analysis since it discusses concepts that lie outside the scope of a typical complex analysis course, such as homotopy and algebraic properties of groups of conformal transformations. In fact, the exposition is non-standard in that the central result, from which most of the material follows, is Cauchy's Independence of Homotopy Theorem (in this regard, I was certainly influenced by A. Vitushkin from whom I took my first complex analysis course and who later became my PhD thesis adviser). Expositions based on the above theorem are hard to come by, and those that I am aware of do not satisfy me, often because of lack of rigour. At the same time, homotopy independence allows one to have a nice clean derivation of Cauchy's Integral Theorem and Cauchy's Integral Formula (see Lecture 10). This is one reason why I have decided to write my own book.

Another instance of a non-standard approach to exposition is the proof of the Fundamental Theorem of Algebra given in Lecture 1. The fact that this important result can be obtained so early in the book and in such an elementary way demonstrates the power of complex numbers and sets the tone for the entire course. The book concludes with a proof of another major milestone, the Riemann Mapping Theorem, which is rarely part of a one-semester undergraduate course. I stress that the exposition is almost entirely self-contained, with only a handful of statements included without proof.

Certainly, the nice extras incorporated in the book as described above come at a price: one has to possess a certain degree of mathematical maturity in order to understand and appreciate them. Namely, the students are required to have done a prerequisite course in real analysis and metric spaces, to which I refer for most facts mentioned without proof. If one does not assume such a course (as is the case at some universities in the US), the instructor may decide to exclude certain topics, e.g., the proof of the Riemann Mapping Theorem, and use the book for teaching a slightly lower-level course. The instructor will find that fitting even such a reduced amount of material in one semester requires a significant effort, so one should not be afraid of running out of content. Alternatively, although this book is primarily aimed at undergraduates, it can be used to teach a graduate course to students who have the right prerequisites.

Before proceeding, I would like to say another word about the exercises. Many of them are of my own making, but over the years I have also collected a good number of interesting unpublished problems informally communicated to me by various people, most of all by A. Vitushkin and E. Chirka. Some of the problems that I learned from them are included in the book, and I am grateful to these two mathematicians for their generous contributions.

Canberra,
June 2017   *Alexander Isaev*

Contents

Lecture 1
Complex Numbers. The Fundamental Theorem of Algebra

The field of complex numbers, or the complex plane, denoted by $\mathbb{C}$, is just the usual Euclidean plane $\mathbb{R}^2$ endowed with the additional operation of multiplication of vectors defined as follows: for (x_1, y_1) and (x_2, y_2) in $\mathbb{R}^2$ let

$$(x_1, y_1) \cdot (x_2, y_2) := (x_1 x_2 - y_1 y_2, x_1 y_2 + x_2 y_1).$$

Notice that if $y_1 = 0$, the above operation is simply the scaling of the vector (x_2, y_2) by x_1.

Consider the standard basis $e_1 := (1, 0)$ and $e_2 := (0, 1)$ in $\mathbb{R}^2$ and write it as $\mathbf{1} := e_1$, $i := e_2$. Then we have

$$(x, y) = xe_1 + ye_2 = x\mathbf{1} + yi,$$

and we abbreviate the latter expression as $x + iy$. Clearly,

$$\mathbf{1} \cdot (x + iy) = x + iy \ \forall x, y \in \mathbb{R},$$

$$i^2 = i \cdot i = -\mathbf{1}.$$

Using the above two rules, it is easy to remember the formula for multiplication by formally multiplying the expressions $(x_1 + iy_1)$ and $(x_2 + iy_2)$:

$$(x_1, y_1) \cdot (x_2, y_2) = (x_1 + iy_1) \cdot (x_2 + iy_2) = x_1 x_2 + ix_1 y_2 + iy_1 x_2 + i^2 y_1 y_2 = (x_1 x_2 - y_1 y_2) + i(x_1 y_2 + x_2 y_1).$$

It is not hard to see that the above multiplication operation, called *complex multiplication*, satisfies the usual field axioms. In particular, as we will explain below, for every non-zero complex number there exists a reciprocal number. In what follows, we omit the symbol $\cdot$ and abbreviate $(x_1 + iy_1) \cdot (x_2 + iy_2)$ as $(x_1 + iy_1)(x_2 + iy_2)$. When we multiply a complex number $(x + iy)$ by itself, we use the symbol $(x + iy)^2$. Similarly, multiplication of n copies of $(x + iy)$ is expressed by the symbol $(x + iy)^n$.

We usually write a complex number as $z = x + iy$, where x is called *the real part of z* (denoted by Re z) and y *the imaginary part of z* (denoted by Im z). If Im $z = 0$, the complex number z is called *real*. We think of the set of all such numbers (namely, the x-axis) as a copy of $\mathbb{R}$ lying in $\mathbb{C}$. Clearly, for real numbers complex multiplication coincides with the usual (real) multiplication, so $\mathbb{C}$ can be viewed as an extension of $\mathbb{R}$.

If $\operatorname{Re} z = 0$, then z is said to be *imaginary*. The set of all imaginary numbers (namely, the y-axis) is geometrically another copy of $\mathbb{R}$ lying in $\mathbb{C}$. Note, however, that this copy is not closed with respect to complex multiplication as the product of two imaginary numbers is real.

We call $x - iy$ *the complex number conjugate to* $z = x + iy$ and denote it by $\bar{z}$. One has $\operatorname{Re} \bar{z} = \operatorname{Re} z$, $\operatorname{Im} \bar{z} = -\operatorname{Im} z$ and

$$\operatorname{Re} z = \frac{z + \bar{z}}{2}, \quad \operatorname{Im} z = -i\frac{z - \bar{z}}{2}.$$

The number $\bar{z}$ should not be confused with *the complex number opposite to* z, which is $-z = -x - iy$.

Define *the modulus, or absolute value, of* z by $|z| := \sqrt{z\bar{z}}$. As $z\bar{z} = x^2 + y^2$, it follows that $|z|$ is the usual Euclidean length of the vector (x, y) in $\mathbb{R}^2$. The modulus has the familiar properties of the modulus of real numbers, e.g.,

$$|z_1 + z_2| \leq |z_1| + |z_2|, \quad |z_1 + z_2| \geq \left||z_1| - |z_2|\right|$$

(check!). The *distance* between any two complex numbers z_1, z_2 is just the Euclidean distance $|z_1 - z_2|$, which turns $\mathbb{C}$ into a metric space. Below, convergence of sequences of complex numbers, limits of $\mathbb{C}$-valued functions, continuity, etc., will be mostly understood with respect to this metric space structure.

Next, for a non-zero z, *the reciprocal of* z, if exists, is the unique complex number w such that $wz = 1$. The reciprocal of z is denoted by either z^{-1} or $\frac{1}{z}$. Using the conjugate number and the modulus, it is easy to see that z^{-1} exists for every non-zero z; in fact, one can compute z^{-1} by scaling the vector $\bar{z}$ as follows:

$$z^{-1} = \frac{1}{|z|^2}\bar{z}.$$

We can now introduce the *ratio of two complex numbers* z_1, z_2, with $z_2 \neq 0$, as

$$\frac{z_1}{z_2} := z_1 z_2^{-1}.$$

Further, in terms of the modulus function, define

$$\Delta(a, r) := \{z \in \mathbb{C} : |z - a| < r\}$$

to be the open disk of radius $0 < r < \infty$ centred at a. We write Δ_r for $\Delta(0, r)$ and denote the *unit disk* Δ_1 simply by Δ. Also, for open disks of zero and infinite radii we set $\Delta(a, 0) := \emptyset$ and $\Delta(a, \infty) := \mathbb{C}$.

Next, one can express any $z \neq 0$ as

$$z = |z|(\cos\varphi, \sin\varphi) = |z|(\cos\varphi + i\sin\varphi),$$

where $0 \le \varphi < 2\pi$ is the angle between the half-line $\mathbb{R}_+ := \{x \in \mathbb{R} : x \ge 0\}$ and the half-line emanating from 0 and passing through z, calculated in the anti-clockwise direction starting with $\mathbb{R}_+$. This number is called *the argument of z* and is denoted by arg z (see Fig. 1.1). Then

$$z = |z|(\cos\arg z + i\sin\arg z) = |z|e^{i\arg z},$$

where for any real number t we set $e^{it} := \cos t + i\sin t$.

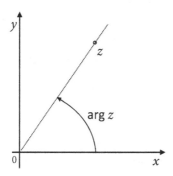

Fig. 1.1

In general, for any $z = x + iy \in \mathbb{C}$ define

$$e^z := e^x(\cos y + i\sin y) = e^x e^{iy}.$$

The function e^z is called *the exponential function*. It has the following familiar property:

Proposition 1.1. *For $z_1, z_2 \in \mathbb{C}$ we have $e^{z_1 + z_2} = e^{z_1} e^{z_2}$.*

Proof. By the definition of the exponential function we write

$$e^{z_1 + z_2} = e^{x_1 + x_2}(\cos(y_1 + y_2) + i\sin(y_1 + y_2)).$$

On the other hand,

$$e^{z_1} e^{z_2} = e^{x_1}(\cos y_1 + i\sin y_1)e^{x_2}(\cos y_2 + i\sin y_2) =$$

$$e^{x_1 + x_2}((\cos y_1 \cos y_2 - \sin y_1 \sin y_2) + i(\cos y_1 \sin y_2 + \sin y_1 \cos y_2)) =$$

$$e^{x_1 + x_2}(\cos(y_1 + y_2) + i\sin(y_1 + y_2)),$$

where we utilised the usual trigonometric identities for cos and sin of sums of arguments. □

Now, fix $z_0 \neq 0$ and consider the multiplication map

$$M_{z_0} : \mathbb{C} \to \mathbb{C}, \quad z \mapsto z_0 z.$$

We have

$$z_0 z = |z_0| e^{i \arg z_0} |z| e^{i \arg z} = |z_0||z| e^{i \arg z_0} e^{i \arg z} \overset{\text{by Proposition 1.1}}{=\!=\!=\!=\!=\!=\!=} |z_0||z| e^{i(\arg z_0 + \arg z)}.$$

Hence, M_{z_0} is simply the composition of the dilation by $|z_0|$ and the anti-clockwise rotation by $\arg z_0$. This is the geometric meaning of complex multiplication.

Note that in general $\arg z_1 + \arg z_2$ may not be equal to $\arg(z_1 z_2)$ owing to the constraint $0 \leq \arg z < 2\pi$ for all non-zero $z \in \mathbb{C}$. However, we have

$$\arg z_1 + \arg z_2 = \arg(z_1 z_2) \,(\text{mod}\, 2\pi).$$

It is therefore convenient to introduce a set-valued analogue of $\arg$ as

$$\text{Arg}\, z := \{\arg z + 2\pi m, \, m \in \mathbb{Z}\}, \, z \neq 0.$$

Indeed, one has

$$\text{Arg}\, z_1 + \text{Arg}\, z_2 = \text{Arg}(z_1 z_2),$$

where the left-hand side is understood as a sum of sets in $\mathbb{R}$ (prove!).

Analogously, for every $z \neq 0$ we define

$$\ln z := \ln |z| + i \arg z \quad \text{and} \quad \text{Ln}\, z := \ln |z| + i \,\text{Arg}\, z.$$

The function $\ln$ is called *the logarithm* and Ln is its set-valued analogue. Note that $e^{\ln z} = z$ for all non-zero z, just as one would expect. Indeed,

$$e^{\ln z} = e^{\ln |z|} (\cos \arg z + i \sin \arg z) = |z|(\cos \arg z + i \sin \arg z) = z.$$

Similarly, one obtains that the set $e^{\text{Ln}\, z}$ in fact consists of the single point z (check!).

Using $\ln z$ and $\text{Ln}\, z$, one can define *powers of complex numbers*. Indeed, for $z, w \in \mathbb{C}$, with $z \neq 0$, let

$$z^w := e^{w \ln z}. \tag{1.1}$$

More generally, for $z \neq 0$ one can introduce a set-valued analogue of the wth power as $e^{w \,\text{Ln}\, z}$. Notice that this set consists of a single point if w is an integer, in which case the definition reduces to the usual definition of integral powers. Indeed, by Proposition 1.1, for all $z \in \mathbb{C}$, $n \in \mathbb{N} = \{1, 2, \ldots\}$ and $m \in \mathbb{Z}$ we have

$$z^n = |z|^n e^{in \arg z} = e^{n \ln z} = e^{n(\ln z + 2\pi i m)}.$$

Next, for $n \in \mathbb{N}$ and any non-zero z the set $e^{\frac{1}{n} \text{Ln}\, z}$ consists of exactly n elements, called *the roots of z of order n*. They are given by the formula

$$|z|^{\frac{1}{n}} e^{i\left(\frac{\arg z}{n} + \frac{2\pi m}{n}\right)}, \quad m = 0, \ldots, n-1. \tag{1.2}$$

These numbers are located on the circle of radius $|z|^{\frac{1}{n}}$ at the vertices of a regular polygon.

Proposition 1.2. *Let $z \neq 0$. Then every $w \in \mathbb{C}$ satisfying $w^n = z$ has the form (1.2).*

Proof. Homework. $\square$

Having introduced basic concepts, we will now demonstrate the power of complex numbers by proving the so-called Fundamental Theorem of Algebra.

Theorem 1.1. (The Fundamental Theorem of Algebra)
   Let

$$P(z) = a_K z^K + a_{K-1} z^{K-1} + \cdots + a_0$$

be a polynomial in z with $a_K \neq 0$, $K \in \mathbb{N}$. Then it has a root in $\mathbb{C}$, that is, a complex number w satisfying $P(w) = 0$. Hence, P factors as $a_K(z - w_1) \cdots (z - w_K)$, where $w_1, \ldots, w_K$ are the roots of P (which are not necessarily all distinct).

Remark 1.1. As the example of the polynomial $z^2 + 1$ shows, even if all the coefficients a_j are real, one cannot expect that P has a real root.

Proof (Theorem 1.1). Let $f(z) := |P(z)|$ and notice that $f : \mathbb{R}^2 \to \mathbb{R}$ is continuous as a composition of continuous functions (check!). Set $s := \inf_{z \in \mathbb{C}} f(z)$.

Lemma 1.1. *For the function f introduced above, there exists a positive number R such that $s = \inf_{z \in \overline{\Delta_R}} f(z)$. Hence, there is $w \in \mathbb{C}$ such that $s = f(w)$.*

Proof. First, we will show that for every $M > 0$ there exists $R > 0$ with $f(z) \geq M$ for $|z| \geq R$. Indeed, assuming that $a_K = 1$ (which can be done without loss of generality), we have

$$f(z) \geq \left| |z|^K - |Q(z)| \right|, \quad \text{where } Q(z) := a_{K-1} z^{K-1} + a_{K-2} z^{K-2} + \cdots + a_0.$$

But

$$|Q(z)| \leq |a_{K-1}||z|^{K-1} + |a_{K-2}||z|^{K-2} + \cdots + |a_0|,$$

which does not exceed $|z|^K/2$ if $|z|$ is sufficiently large. Therefore, $f(z) \geq |z|^K/2$ for all sufficiently large $|z|$, and thus for any $M > 0$ there exists $R > 0$ such that $f(z) \geq M$ for $|z| \geq R$ as required.

Choosing $M > s$, we then see that $s = \inf_{z \in \overline{\Delta_R}} f(z)$. As f is continuous and $\overline{\Delta_R}$ is compact, the infimum is attained at some point $w \in \overline{\Delta_R}$. $\square$

We shall continue with the proof of the theorem. Let $w \in \mathbb{C}$ be a point as in Lemma 1.1. We will now show that $s = 0$, which will mean that w is a root of P. Assume that $s > 0$. Then the entire image of the complex plane $\mathbb{C}$ under P lies outside the non-empty disk Δ_s as illustrated in Fig.1.2 (recall that $s = |P(w)|$). Notice that in the figure the strokes indicate *the complement* to $P(\mathbb{C})$; in what follows we shall often mark sets of interest by drawing strokes in their complements.

Write $z = w + (z - w)$ and express $P(z)$ via powers of $z - w$ by expanding every z^ℓ as

$$z^\ell = \sum_{j=0}^{\ell} \binom{\ell}{j} w^j (z - w)^{\ell - j}.$$

We thus obtain

$$P(z) = P(w) + b_k(z - w)^k + \cdots + b_K(z - w)^K,$$

with $b_k \neq 0$.

Lemma 1.2. *For any* $0 \leq \varphi < 2\pi$ *there exists* $z_0 \in \mathbb{C}$ *arbitrarily close to* w *with* $\arg(b_k(z_0 - w)^k) = \varphi$.

Proof. It suffices to prove the lemma for $b_k = 1$ (explain!). Set $z_0 = w + \zeta$, where $|\zeta|$ is arbitrarily small and non-zero. If $\psi := \arg \zeta$, then $k\psi = \arg \zeta^k \pmod{2\pi}$, and the lemma follows by choosing ζ with $\psi = \varphi/k$. $\square$

We shall now finalise the proof of the theorem. For z sufficiently close to w we have

$$|b_{k+1}(z - w)^{k+1} + \cdots + b_K(z - w)^K| \leq |b_{k+1}||z - w|^{k+1} + \cdots + |b_K||z - w|^K \leq$$
$$\tfrac{1}{2}|b_k||z - w|^k.$$

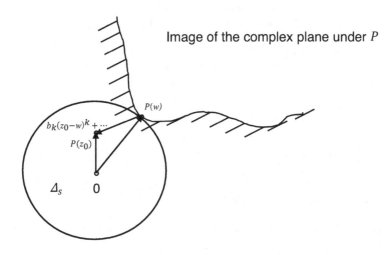

Image of the complex plane under P

Fig. 1.2

By choosing an appropriate φ, in accordance with Lemma 1.2 we can now ensure that for some z_0 arbitrarily close to w the vector

$$W := P(w) + b_k(z_0 - w)^k + b_{k+1}(z_0 - w)^{k+1} + \cdots + b_K(z_0 - w)^K$$

lies in Δ_s. Indeed, thinking of W as the sum of the vectors $P(w)$ and $b_k(z_0 - w)^k + b_{k+1}(z_0 - w)^{k+1} + \cdots + b_K(z_0 - w)^K$, we can guarantee that $|W| < s$ (see Fig. 1.2, where the latter vector is shown as emanating from $P(w)$). But, since $W = P(z_0)$, this conclusion contradicts the fact that the image of $\mathbb{C}$ under P is outside Δ_s. As the contradiction came from the assumption $s > 0$, we have $s = 0$. Hence, w is indeed a root of P, and the proof of the theorem is complete. □

Exercises

1.1. Compute the values of the following expressions:

$$\text{(i) } i(3 + 4i),$$
$$\text{(ii) } (\sqrt{2} + i\sqrt{3})(\sqrt{2} - \sqrt{3}i),$$
$$\text{(iii) } (\sqrt{2} + i\sqrt{3})^3.$$

1.2. For each of the following numbers, find its real part, imaginary part, complex conjugate number, opposite number, modulus, reciprocal number, and argument:

$$\text{(i) } 1 + i,$$
$$\text{(ii) } 2 - 2i,$$
$$\text{(iii) } \sqrt{3} - i,$$
$$\text{(iv) } -3 - 3i\sqrt{3}.$$

In addition, for every pair of the above numbers, compute the ratio.

1.3. For each of the following numbers, compute the value of e^z:

$$\text{(i) } 1 + \frac{\pi}{2}i,$$
$$\text{(ii) } 2 - \frac{3i\pi}{4},$$
$$\text{(iii) } \sqrt{3} - \frac{\pi i}{6},$$
$$\text{(iv) } -3 - i\frac{\pi}{3}.$$

1.4. For each of the numbers from Exercise 1.2, compute the value of $\ln z$.

1.5. Compute i^i, i^{-i}, $(-i)^i$, $(-i)^{-i}$.

1.6. Find all roots of 3 of order 3.

1.7. Let $z_1, \ldots, z_n$ be non-zero complex numbers such that

$$\sum_{k=1}^{n} \frac{1}{z_k} = 0.$$

Prove that for any line passing through the origin, the numbers $z_1, \ldots, z_n$ cannot all lie in a half-plane on either side of the line.

1.8. Prove that if z_1, z_2, z_3 are complex numbers satisfying $|z_1| = |z_2| = |z_3| = 1$ and $z_1 + z_2 + z_3 = 0$, then they split the unit circle $S^1 := \{z \in \mathbb{C} : |z| = 1\}$ into three arcs of equal sizes.

1.9. Show that if $|z| = 1$ and $z \neq \pm 1$, then

$$\text{either } \arg\left(\frac{z-1}{z+1}\right) = \frac{\pi}{2}, \text{ or } \arg\left(\frac{z-1}{z+1}\right) = \frac{3\pi}{2}.$$

1.10. Find a mistake in the following reasoning:

$$z^2 = (-z)^2 \Rightarrow \text{Arg}(z^2) = \text{Arg}((-z)^2)) \Rightarrow 2\,\text{Arg}\,z = 2\,\text{Arg}(-z) \Rightarrow$$
$$\text{Arg}\,z = \text{Arg}(-z) \Rightarrow \text{Arg}\,1 = \text{Arg}(-1).$$

1.11. Let $\Phi : \mathbb{R} \to \mathbb{R}$ be a continuous function satisfying $\Phi(x) \in \text{Arg}(e^{ix})$ for all $x \in \mathbb{R}$. Prove that there exists $k \in \mathbb{Z}$ such that

$$\Phi(x) = x + 2\pi k \; \forall x \in \mathbb{R}.$$

1.12. Prove that the set of values of the exponential function on $\mathbb{C} \setminus \overline{A}$ is $\mathbb{C} \setminus \{0\}$.

1.13. Let $f : \mathbb{C} \to \mathbb{C}$ be a continuous function. Assume that there exist $K \in \mathbb{N}$ and complex numbers $a_0, \ldots, a_{K-1}$ for which

$$f^K + a_{K-1} f^{K-1} + \cdots + a_1 f + a_0 \equiv 0.$$

Prove that $f \equiv \text{const}$.

Lecture 2
$\mathbb{R}$- and $\mathbb{C}$-Differentiability

Let $z_0 = x_0 + iy_0 = (x_0, y_0)$ be a point in $\mathbb{C}$ and f a function defined on a neighbourhood of z_0 (e.g., on an open disk $\Delta(z_0, r)$ for some $r > 0$) with values in $\mathbb{C}$. Write $f(z) = \mathrm{Re}\, f(z) + i\,\mathrm{Im}\, f(z) = u(z) + iv(z) = u(x,y) + iv(x,y)$.

Definition 2.1. The function f is called $\mathbb{R}$-*differentiable at* z_0 if there exist real numbers a, b, c, d such that

$$u(x,y) = u(x_0, y_0) + a\Delta x + b\Delta y + o(\Delta x, \Delta y),$$

$$v(x,y) = v(x_0, y_0) + c\Delta x + d\Delta y + o(\Delta x, \Delta y),$$

where $\Delta x := x - x_0$, $\Delta y := y - y_0$ and $o(\Delta x, \Delta y)$ denotes any real-valued function with the property

$$\frac{o(\Delta x, \Delta y)}{\sqrt{(\Delta x)^2 + (\Delta y)^2}} \to 0 \text{ as } (\Delta x)^2 + (\Delta y)^2 \to 0.$$

Here a, b, c, d are determined uniquely and are in fact the corresponding *first-order partial derivatives of u and v at (x_0, y_0)*:

$$a = \frac{\partial u}{\partial x}(x_0, y_0), \quad b = \frac{\partial u}{\partial y}(x_0, y_0), \quad c = \frac{\partial v}{\partial x}(x_0, y_0), \quad d = \frac{\partial v}{\partial y}(x_0, y_0).$$

We will now introduce the concept of complex differentiability and compare it with that of real differentiability as defined above.

Definition 2.2. The function f is said to be $\mathbb{C}$-*differentiable at* z_0 if there exists the limit

$$\lim_{\Delta z \to 0} \frac{f(z_0 + \Delta z) - f(z_0)}{\Delta z},$$

that is, one can find $A \in \mathbb{C}$ such that for any $\varepsilon > 0$ there is a sufficiently small $\delta > 0$ with the property

$$\left| \frac{f(z_0 + \Delta z) - f(z_0)}{\Delta z} - A \right| < \varepsilon$$

9

if $0 < |\Delta z| < \delta$. In this case, the complex number A, i.e., the value of the limit, is called *the derivative of f at z_0* and is denoted by $f'(z_0)$.

The usual rules from real analysis for computing the derivatives of sums, products, quotients and compositions of ℂ-differentiable functions work here (check!).

Definition 2.2 is equivalent to saying that for some $A \in \mathbb{C}$ the function f is represented as

$$f(z) = f(z_0) + A\Delta z + o(\Delta z), \tag{2.1}$$

where $\Delta z := z - z_0$ and $o(\Delta z)$ denotes any complex-valued function with the property

$$\frac{o(\Delta z)}{\Delta z} \to 0 \text{ as } \Delta z \to 0$$

(note that one can write $o(\Delta z)$ instead of $o(\Delta x, \Delta y)$ in Definition 2.1). By separating the real and imaginary parts in identity (2.1), we see that it is equivalent to the following pair of real identities:

$$u(x,y) = u(x_0,y_0) + \operatorname{Re} A\,\Delta x - \operatorname{Im} A\,\Delta y + o(\Delta x, \Delta y),$$

$$v(x,y) = v(x_0,y_0) + \operatorname{Im} A\,\Delta x + \operatorname{Re} A\,\Delta y + o(\Delta x, \Delta y).$$

Comparing these identities with Definition 2.1 and taking into account that the constants a, b, c, d are chosen uniquely, we obtain:

Theorem 2.1. *The function f is ℂ-differentiable at $z_0 = x_0 + iy_0$ if and only if it is ℝ-differentiable at z_0 and the first-order partial derivatives of u and v at (x_0, y_0) satisfy the relations*

$$\frac{\partial u}{\partial x}(x_0,y_0) = \frac{\partial v}{\partial y}(x_0,y_0), \quad \frac{\partial u}{\partial y}(x_0,y_0) = -\frac{\partial v}{\partial x}(x_0,y_0). \tag{2.2}$$

Relations (2.2) are called *the Cauchy-Riemann equations* or simply *the CR-equations*. We will now rewrite them in a different form.

Definition 2.3. Let f be ℝ-differentiable at z_0. Set

$$\frac{\partial f}{\partial z}(z_0) := \frac{1}{2}\left(\frac{\partial f}{\partial x}(x_0,y_0) - i\frac{\partial f}{\partial y}(x_0,y_0)\right), \quad \frac{\partial f}{\partial \bar{z}}(z_0) := \frac{1}{2}\left(\frac{\partial f}{\partial x}(x_0,y_0) + i\frac{\partial f}{\partial y}(x_0,y_0)\right),$$

where

$$\frac{\partial f}{\partial x}(x_0,y_0) := \frac{\partial u}{\partial x}(x_0,y_0) + i\frac{\partial v}{\partial x}(x_0,y_0), \quad \frac{\partial f}{\partial y}(x_0,y_0) := \frac{\partial u}{\partial y}(x_0,y_0) + i\frac{\partial v}{\partial y}(x_0,y_0).$$

Now, multiplying the second identity from Definition 2.1 by i, adding it to the first one and substituting

$$\Delta x = \frac{\Delta z + \overline{\Delta z}}{2}, \quad \Delta y = \frac{\Delta z - \overline{\Delta z}}{2i},$$

we obtain

$$f(z) = f(z_0) + \left(\frac{\partial u}{\partial x}(x_0,y_0) + i\frac{\partial v}{\partial x}(x_0,y_0) \right) \frac{\Delta z + \overline{\Delta z}}{2} +$$

$$\left(\frac{\partial u}{\partial y}(x_0,y_0) + i\frac{\partial v}{\partial y}(x_0,y_0) \right) \frac{\Delta z - \overline{\Delta z}}{2i} + o(\Delta z) =$$

$$f(z_0) + \frac{1}{2} \left(\frac{\partial f}{\partial x}(x_0,y_0) - i\frac{\partial f}{\partial y}(x_0,y_0) \right) \Delta z +$$

$$\frac{1}{2} \left(\frac{\partial f}{\partial x}(x_0,y_0) + i\frac{\partial f}{\partial y}(x_0,y_0) \right) \overline{\Delta z} + o(\Delta z) =$$

$$f(z_0) + \frac{\partial f}{\partial z}(z_0)\Delta z + \frac{\partial f}{\partial \bar{z}}(z_0)\overline{\Delta z} + o(\Delta z).$$

Comparing the above formula with (2.1), we see:

Theorem 2.2. *The function f is* ℂ*-differentiable at* z_0 *if and only if it is* ℝ*-differentiable at* z_0 *and* $\frac{\partial f}{\partial \bar{z}}(z_0) = 0$. *In this case* $f'(z_0) = \frac{\partial f}{\partial z}(z_0)$.

Proof. Homework. (Hint: use, e.g., Exercise 2.4.) □

Remark 2.1. The fact that the CR-equations from Theorem 2.1 are equivalent to the condition $\frac{\partial f}{\partial \bar{z}}(z_0) = 0$ can be also seen directly from Definition 2.3 by separating the real and imaginary parts of $\frac{\partial f}{\partial \bar{z}}(z_0)$. Indeed, we have

$$\frac{\partial f}{\partial \bar{z}}(z_0) = \frac{1}{2} \left(\frac{\partial u}{\partial x}(x_0,y_0) - \frac{\partial v}{\partial y}(x_0,y_0) \right) + \frac{i}{2} \left(\frac{\partial v}{\partial x}(x_0,y_0) + \frac{\partial u}{\partial y}(x_0,y_0) \right),$$

hence the complex identity $\frac{\partial f}{\partial \bar{z}}(z_0) = 0$ is equivalent to the pair of real identities in (2.2).

The "formal derivatives" $\frac{\partial f}{\partial z}$ and $\frac{\partial f}{\partial \bar{z}}$ can be often treated as "usual" derivatives. For example, consider a complex-valued polynomial in x, y

$$P(x,y) = \sum_{0 \le \ell,m \le K} a_{\ell m} x^\ell y^m, \ a_{\ell m} \in \mathbb{C}.$$

Substituting

$$x = \frac{z + \bar{z}}{2}, \quad y = \frac{z - \bar{z}}{2i},$$

we express P via z, $\bar{z}$:

$$P = \sum_{0 \le \ell,m \le K} b_{\ell m} z^\ell \bar{z}^m$$

for some $b_{\ell m} \in \mathbb{C}$.

Proposition 2.1. *For all $z_0 \in \mathbb{C}$ we have*

$$\frac{\partial P}{\partial z}(z_0) = \sum_{0 \leq \ell, m \leq K} \ell b_{\ell m} z_0^{\ell-1} \bar{z}_0^m, \quad \frac{\partial P}{\partial \bar{z}}(z_0) = \sum_{0 \leq \ell, m \leq K} m b_{\ell m} z_0^{\ell} \bar{z}_0^{m-1}.$$

Proof. It suffices to prove the proposition for any monomial $Q := z^\ell \bar{z}^m$. Since $Q = (x + iy)^\ell (x - iy)^m$, for $z_0 = x_0 + iy_0$ we calculate

$$\frac{\partial Q}{\partial x}(x_0, y_0) = \ell(x_0 + iy_0)^{\ell-1}(x_0 - iy_0)^m + m(x_0 + iy_0)^\ell (x_0 - iy_0)^{m-1} =$$
$$\ell z_0^{\ell-1} \bar{z}_0^m + m z_0^\ell \bar{z}_0^{m-1},$$

$$\frac{\partial Q}{\partial y}(x_0, y_0) = i\ell(x_0 + iy_0)^{\ell-1}(x_0 - iy_0)^m - im(x_0 + iy_0)^\ell (x_0 - iy_0)^{m-1} =$$
$$i\ell z_0^{\ell-1} \bar{z}_0^m - im z_0^\ell \bar{z}_0^{m-1}$$

(check!). Hence,

$$\frac{\partial Q}{\partial z}(z_0) = \frac{1}{2}\left(\frac{\partial Q}{\partial x}(x_0, y_0) - i\frac{\partial Q}{\partial y}(x_0, y_0)\right) = \ell z_0^{\ell-1} \bar{z}_0^m,$$
$$\frac{\partial Q}{\partial \bar{z}}(z_0) = \frac{1}{2}\left(\frac{\partial Q}{\partial x}(x_0, y_0) + i\frac{\partial Q}{\partial y}(x_0, y_0)\right) = m z_0^\ell \bar{z}_0^{m-1},$$

which completes the proof. □

We will be mostly interested in the ℂ-differentiability of functions on open subsets of the complex plane.

Definition 2.4. *Let $D \subset \mathbb{C}$ be an open subset. A function $f : D \to \mathbb{C}$ is said to be* holomorphic on D *if f is ℂ-differentiable at every point of D. All functions holomorphic on D form a vector space over ℂ, which we denote by $H(D)$. Functions in $H(\mathbb{C})$ (i.e., those holomorphic on all of ℂ) are called* entire.

We have the usual facts:

(1) if $f, g \in H(D)$, then $f + g \in H(D)$ and $fg \in H(D)$;
(2) if $f, g \in H(D)$ and g does not vanish at any point of D, then $f/g \in H(D)$;
(3) if $f \in H(D)$, $g \in H(G)$, $f(D) \subset G$, then $g \circ f \in H(D)$.

The proofs are identical to those for functions differentiable on open subsets of ℝ from real analysis (check!).

Usually, in what follows D will be a *domain*, i.e., an open *connected* subset of ℂ, where connectedness is understood as the existence, for any $z, w \in D$, of *a path in D joining z and w*, that is, of a continuous map $\gamma : [0, 1] \to D$ with $\gamma(0) = z$, $\gamma(1) = w$ (in this case, we sometimes say that z is the *initial point* of γ and w is its *terminal point*). Strictly speaking, the above property is called *path-connectedness* but for open subsets of $\mathbb{R}^n$ (in fact, for open subsets of any locally path-connected topological space) it is equivalent to connectedness, i.e., to the non-existence of a non-trivial subset that is both open and closed (check!).

In what follows, we will be often interested in extending a function $f \in H(D)$ holomorphically beyond the domain D. Namely, for a domain $G \supset D$ we say that f [*holomorphically*] *extends* [*to* G], or that f *can be* [*holomorphically*] *extended* [*to* G], if there exists $F \in H(G)$ such that $F(z) = f(z)$ for all $z \in D$. In this case F is called *a holomorphic extension of* f [*to* G] and we also say that f *extends to a function holomorphic on* G, or that f *can be extended to a function holomorphic on* G.

Example 2.1. Let $f(z) := z = x + iy$. Here $u = x$, $v = y$, hence

$$\frac{\partial u}{\partial x} = \frac{\partial v}{\partial y} \equiv 1, \quad \frac{\partial u}{\partial y} = \frac{\partial v}{\partial x} \equiv 0.$$

Thus, we see that f is an entire function. It then follows that every polynomial $P(z) = a_K z^K + a_{K-1} z^{K-1} + \cdots + a_0$ in z is an entire function.

Example 2.2. Let $f(z) := e^z = e^x(\cos y + i \sin y)$. Here $u = e^x \cos y$, $v = e^x \sin y$, hence

$$\frac{\partial u}{\partial x} = \frac{\partial v}{\partial y} = e^x \cos y, \quad \frac{\partial u}{\partial y} = -\frac{\partial v}{\partial x} = -e^x \sin y.$$

Therefore, f is an entire function. Also, for any $z_0 \in \mathbb{C}$ we have

$$f'(z_0) = \frac{\partial f}{\partial z}(z_0) = \frac{1}{2}\left(\frac{\partial f}{\partial x}(x_0, y_0) - i\frac{\partial f}{\partial y}(x_0, y_0)\right) = \frac{1}{2}(e^{z_0} - i(ie^{z_0})) = e^{z_0}$$

as expected. It then follows that *the basic trigonometric functions*

$$\cos z := \frac{e^{iz} + e^{-iz}}{2}, \quad \sin z := \frac{e^{iz} - e^{-iz}}{2i}$$

are entire as well.

Example 2.3. Let $f(z) := \bar{z} = x - iy$. Here $u = x$, $v = -y$, hence

$$\frac{\partial u}{\partial x} \equiv 1, \quad \frac{\partial v}{\partial y} \equiv -1, \quad \frac{\partial u}{\partial y} = \frac{\partial v}{\partial x} \equiv 0.$$

Therefore, f is not ℂ-differentiable at any point of ℂ.

Example 2.4. Let $f(z) := \bar{z}^2 = x^2 - y^2 - 2ixy$. Here $u = x^2 - y^2$, $v = -2xy$, hence

$$\frac{\partial u}{\partial x} = 2x, \quad \frac{\partial v}{\partial y} = -2x, \quad \frac{\partial u}{\partial y} = -2y, \quad \frac{\partial v}{\partial x} = -2y.$$

Therefore, f is only ℂ-differentiable at the origin. In particular, f is not holomorphic on any domain in ℂ.

Notice also that as a consequence of Proposition 2.1 we have:

Corollary 2.1. *The polynomial P from Proposition 2.1 is an entire function if and only if* $b_{\ell,m} = 0$ *for all* $m > 0$. *In this case the derivative* $P' = \dfrac{\partial P}{\partial z}$ *is given by formal differentiation with respect to z.*

Proof. Homework. (Hint: use exercise 2.10.) $\square$

Exercises

2.1. Let f be $\mathbb{R}$-differentiable at z_0. Prove that the Jacobian of f at z_0 is equal to

$$\left| \frac{\partial f}{\partial z}(z_0) \right|^2 - \left| \frac{\partial f}{\partial \bar{z}}(z_0) \right|^2.$$

2.2. Let f be $\mathbb{R}$-differentiable at z_0. Prove that the *limit set* of the ratio

$$g(z) := \frac{f(z) - f(z_0)}{z - z_0}, \quad z \neq z_0,$$

at z_0 is the circle centred at the point $\dfrac{\partial f}{\partial z}(z_0)$ of radius $\left| \dfrac{\partial f}{\partial \bar{z}}(z_0) \right|$. Here the limit set consists of all *limit points* of g at z_0, i.e., of all complex numbers A for which there is a sequence $\{z_n\}$ not including the point z_0, with $|z_n - z_0| \to 0$ as $n \to \infty$, such that $|g(z_n) - A| \to 0$.

2.3. Let f be continuously differentiable on $\mathbb{C}$. Suppose that f preserves distances, i.e.,

$$|f(z_1) - f(z_2)| = |z_1 - z_2| \ \forall z_1, z_2 \in \mathbb{C}.$$

Prove that either $f(z) = e^{i\alpha} z + a$, or $f(z) = e^{i\alpha}\bar{z} + a$ for some $a \in \mathbb{C}$ and $\alpha \in \mathbb{R}$. (Hint: use Exercise 2.2.)

2.4. Assume that f is defined on a neighbourhood of a point z_0 and in this neighbourhood for some $A, B \in \mathbb{C}$ the following holds:

$$f(z) = f(z_0) + A\Delta z + B\overline{\Delta z} + o(\Delta z),$$

where $\Delta z := z - z_0$. Prove that f is $\mathbb{R}$-differentiable at z_0 with $A = \dfrac{\partial f}{\partial z}(z_0)$ and $B = \dfrac{\partial f}{\partial \bar{z}}(z_0)$.

2.5. Suppose that $f = u + iv$ is defined on a neighbourhood of 0 and is continuous at 0. Assume that all first-order partial derivatives of u and v exist at 0 and satisfy the CR-equations at 0. Does it follow that f is $\mathbb{C}$-differentiable at the origin? Prove your conclusion.

2.6. For each of the following functions, find all points at which it is $\mathbb{C}$-differentiable:

$$\text{(i) } z^2|z|^4,$$
$$\text{(ii) } (\text{Re } z)^4,$$
$$\text{(iii) } \sin(\text{Im } z).$$

2.7. Let $f = u + iv$ be $\mathbb{C}$-differentiable at z_0, with $f'(z_0) \neq 0$, and continuously differentiable on a neighbourhood of z_0. Prove that the angle between the level sets of u and v at z_0 (that is, any of the four angles between the tangent lines to the level sets at z_0) is equal to $\pi/2$. (Hint: write the tangent lines to the level sets at z_0 via the first-order partial derivatives of u and v at z_0.)

2.8. Suppose that a function f is $\mathbb{C}$-differentiable at a point z_0. Prove that the function

$$g(z) := \overline{f(\bar{z})}$$

is $\mathbb{C}$-differentiable at the point $\bar{z}_0$ and $g'(\bar{z}_0) = \overline{f'(z_0)}$.

2.9. Suppose that a function f is $\mathbb{C}$-differentiable at a point z_0 and $f'(z_0) \neq 0$. Show that for any disk $\Delta(z_0, r)$ on which f is defined the set $f(\Delta(z_0, r))$ cannot lie in a half-plane on either side of any line passing through $f(z_0)$.

2.10. Show that for a polynomial

$$P(z, \bar{z}) = \sum_{0 \le \ell, m \le K} b_{\ell m} z^\ell \bar{z}^m, \ b_{\ell m} \in \mathbb{C},$$

in $z, \bar{z}$ one has $P \equiv 0$ on an open subset of $\mathbb{C}$ if and only if $b_{\ell m} = 0$ for all $\ell, m \in \{0, \dots, K\}$. (Hint: argue by induction using Proposition 2.1.)

2.11. Suppose that for a polynomial

$$P(z, \bar{z}) = \sum_{0 \le \ell, m \le K} b_{\ell m} z^\ell \bar{z}^m, \ b_{\ell m} \in \mathbb{C},$$

in $z, \bar{z}$ we have $b_{\ell m} \neq 0$ for some $0 \le \ell \le K$ and some $0 < m \le K$. Prove that the set of points at which P is $\mathbb{C}$-differentiable is nowhere dense in $\mathbb{C}$. (Hint: use Exercise 2.10.)

2.12. Find the derivatives of $\sin z$ and $\cos z$ at an arbitrary point of $\mathbb{C}$.

2.13. Prove that $|\cos z|$ and $|\sin z|$ are not bounded on $\mathbb{C}$.

2.14. How many zeroes does the entire function $2 + \sin z$ have in $\mathbb{C}$? Find all the zeroes.

2.15. Using the power series expansions of $e^x, \cos y, \sin y$, find a power series expansion with centre 0 for each of the functions $e^z, \cos z, \sin z$, i.e., represent each of these functions in the form

$$\sum_{n=0}^{\infty} c_n z^n \; \forall z \in \mathbb{C},$$

where $c_n \in \mathbb{C}$, $n = 0, 1, 2, \ldots$. (Hint: substitute the corresponding series into the expression $e^x \cos y + i e^x \sin y$ using arithmetic operations with absolutely convergent series and the possibility of re-arranging their terms.)

2.16. Prove that the function $\dfrac{\sin z}{z}$ is holomorphic on $\mathbb{C} \setminus \{0\}$ and that it holomorphically extends to $\mathbb{C}$.

2.17. Let

$$f(z) := \begin{cases} z^2 \sin \dfrac{1}{z} & \text{if } z \neq 0, \\[2mm] 0 & \text{if } z = 0. \end{cases}$$

Is this function $\mathbb{C}$-differentiable at 0? Prove your conclusion.

2.18. Let $f(z) = u(x) + iv(y)$ be an entire function. Prove that $f(z) = az + b$, where $a \in \mathbb{R}$, $b \in \mathbb{C}$.

2.19. Let $a, b, c \in \mathbb{C}$. Write the quadratic polynomial $P(x, y) = ax^2 + bxy + cy^2$ in x, y via z and $\bar{z}$ and find necessary and sufficient conditions on a, b, c for P to be an entire function.

2.20. Find all entire functions f such that $\operatorname{Re} f(z) = x^2 - y^2$.

2.21. Find an entire functions f such that $\operatorname{Re} f(z) = x^2 - y^2 + xy$, $f(0) = 0$.

2.22. Find an entire functions f such that $\operatorname{Re} f(z) = e^x(x \cos y - y \sin y)$, $f(0) = 0$.

2.23. Let $g : [0, 1] \to \mathbb{C}$ be a continuous function. For all $z \in \mathbb{C} \setminus [0, 1]$ define

$$f(z) := \frac{1}{2\pi i} \int_0^1 \frac{g(t)}{t - z} \, dt, \tag{2.3}$$

where the integral is understood by separating the real and imaginary parts of the integrand. Prove that $f \in H(\mathbb{C} \setminus [0, 1])$.

2.24. Construct an example showing that the Mean Value Theorem for $\mathbb{C}$-valued functions does not hold. Namely, find a differentiable function $f : [0, 1] \to \mathbb{C}$ such that $f'(t) \neq f(1) - f(0)$ for all $t \in (0, 1)$.

2.25. Prove the following variant of the Mean Value Theorem for $\mathbb{C}$-valued functions: if $f : [0, 1] \to \mathbb{C}$ is a continuously differentiable function, then the difference $f(1) - f(0)$ lies in the closure of the convex hull of the set $\{f'(t) : t \in [0, 1]\}$. (Hint: use the Newton-Leibniz formula for $\operatorname{Re} f$ and $\operatorname{Im} f$.)

2.26. Let P be a polynomial in z of positive degree and S the smallest convex polygon containing the roots of P. Prove that all roots of P' lie in S. (Hint: use Exercise 1.7.)

Lecture 3
The Stereographic Projection. Conformal Maps.
The Open Mapping Theorem

In the previous lecture we introduced functions holomorphic on domains in $\mathbb{C}$. Here we will allow domains to include "the infinity" and look at the so-called *conformal maps* on such extended domains. As we shall see, maps of this kind are analogous to one-to-one holomorphic functions.

First of all, we shall add "the infinity" to $\mathbb{C}$ in the following way. Consider $\mathbb{R}^3$ with coordinates (x, y, w), the sphere $S^2 \subset \mathbb{R}^3$ with centre $(0, 0, 1/2)$ of radius $1/2$ and think of the complex plane $\mathbb{C}$ as the subspace $\{w = 0\}$ of $\mathbb{R}^3$. Let $N := (0, 0, 1)$ be the "North Pole" of the sphere. We will now look at the 1-to-1 map Π between $S^2 \setminus \{N\}$ and $\mathbb{C}$ shown in Fig. 3.1.

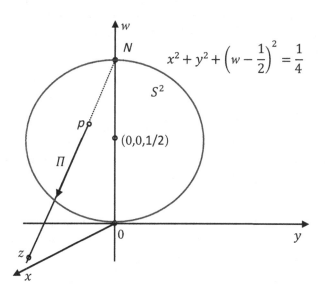

Fig. 3.1

It takes a point p of the sphere distinct from N to the point $z \in \mathbb{C}$ and is called *the stereographic projection*. Notice that $\Pi(0) = 0$.

Let us find explicit formulas for Π and Π^{-1}. First of all, the parametric equation of the line through the points $(0,0,1)$ and $(x,y,0)$ is $(tx, ty, 1-t), t \in \mathbb{R}$. It intersects S^2 if

$$(tx)^2 + (ty)^2 + \left(1 - t - \frac{1}{2}\right)^2 = \frac{1}{4},$$

or, equivalently, if

$$(tx)^2 + (ty)^2 + t^2 - t = 0,$$

or, equivalently, if either $t = 0$ (the point N) or

$$t = \frac{1}{1 + x^2 + y^2} = \frac{1}{1 + |z|^2} \text{ (the point } p\text{)}.$$

Therefore, we have

$$\Pi^{-1}(x,y) = \left(\frac{x}{1 + |z|^2}, \frac{y}{1 + |z|^2}, \frac{|z|^2}{1 + |z|^2}\right).$$

It then follows that

$$\Pi(x, y, w) = \left(\frac{x}{1-w}, \frac{y}{1-w}\right).$$

The formulas for Π and Π^{-1} yield, in particular, that Π is a homeomorphism between $S^2 \setminus \{N\}$ and $\mathbb{C}$. Therefore, the entire sphere S^2 can be thought of as a compact topological space obtained from $\mathbb{C}$ by adding a single point, namely, the point N. Constructions of this kind are called *one-point compactifications* of topological spaces. We will think of the additional point N as *the infinity, or infinite point, added to* $\mathbb{C}$ and denote it by ∞. Instead of S^2 we usually write $\overline{\mathbb{C}}$ and call it *the Riemann sphere* or *the extended complex plane* thus emphasising the fact that S^2 will always be considered together with the stereographic projection.

We will often study maps defined on *domains in* $\overline{\mathbb{C}}$, where a domain is an open connected subset of $\overline{\mathbb{C}}$. Here the openness of a set $D \subset \overline{\mathbb{C}}$ means that it can be represented as the intersection of an open subset of $\mathbb{R}^3$ with S^2. In particular, if D is open and contains ∞, then $\Pi(D \setminus \{N\})$ contains the complement to Δ_R for some $R > 0$. Also, the connectedness of an open set $D \subset \overline{\mathbb{C}}$ is understood as the existence, for every pair of points $p_1, p_2 \in D$, of *a path in D joining p_1 and p_2*, that is, of a continuous map $\gamma : [0,1] \to \mathbb{R}^3$ with $\gamma([0,1]) \subset D$ and $\gamma(0) = p_1$, $\gamma(1) = p_2$. If D is a domain in $\mathbb{C}$, we will usually think of it as a domain in $\overline{\mathbb{C}}$ by identifying D with $\Pi^{-1}(D)$; in particular, we identify $\mathbb{C}$ and $\overline{\mathbb{C}} \setminus \{\infty\}$ without mentioning this explicitly. In the same way, we often identify $z \in \mathbb{C}$ with $\Pi^{-1}(z) \in \overline{\mathbb{C}}$ and a map $f : S \to T$ between subsets of $\mathbb{C}$ with the map $\Pi^{-1} \circ f \circ \Pi$ between the subsets $\Pi^{-1}(S), \Pi^{-1}(T)$ of $\overline{\mathbb{C}}$.

In order to introduce conformal maps, we will first define a special class of paths and angles between them. Recall first that a path in $\mathbb{C}$ is a map from $[0,1]$ to $\mathbb{R}^2 = \mathbb{C}$; we usually write a path γ as $\gamma(t) = (\gamma_1(t), \gamma_2(t)) = \gamma_1(t) + i\gamma_2(t)$.

Definition 3.1. A path γ in $\mathbb{C}$ is called *smooth at* $t_0 \in (0,1)$ if γ is differentiable at t_0 (i.e., each of the functions γ_1, γ_2 is differentiable at t_0) and its derivative $\gamma'(t_0) = (\gamma_1'(t_0), \gamma_2'(t_0)) = \gamma_1'(t_0) + i\gamma_2'(t_0)$ at t_0 is non-zero.

Definition 3.2. Let γ and $\tilde{\gamma}$ be two paths in $\mathbb{C}$ with $\gamma(t_0) = \tilde{\gamma}(\tilde{t}_0) = z_0$. Assume that γ is smooth at t_0 and $\tilde{\gamma}$ is smooth at $\tilde{t}_0$. *The oriented angle between* γ *and* $\tilde{\gamma}$ *at* z_0 *relative to* t_0, $\tilde{t}_0$ is *the oriented angle between the vectors* $\gamma'(t_0)$ *and* $\tilde{\gamma}'(\tilde{t}_0)$, that is, the angle between $\gamma'(t_0)$ and $\tilde{\gamma}'(\tilde{t}_0)$ calculated beginning with $\gamma'(t_0)$ in the anti-clockwise direction.

Notice that the angle between γ and $\tilde{\gamma}$ in most cases differs from that between $\tilde{\gamma}$ and γ, with the sum of these two angles being 2π. In Fig. 3.2, $\gamma'(t_0)$ and $\tilde{\gamma}'(\tilde{t}_0)$ are shown as emanating from the point z_0 in order to emphasise that they are direction vectors of *the tangent lines to the paths at* z_0 *relative to* t_0, $\tilde{t}_0$. In what follows, for a path γ smooth at a point t_0, we refer to the derivative $\gamma'(t_0)$ as *the tangent vector to* γ *at* t_0.

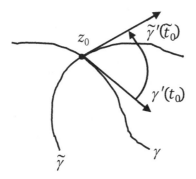

Fig. 3.2

Definition 3.3. Let $z_0 = x_0 + iy_0$ and f be a function defined on a neighbourhood $U \subset \mathbb{C}$ of z_0 with values in $\mathbb{C}$. We say that f is *conformal at* z_0 if

(1) f is $\mathbb{R}$-differentiable at z_0;
(2) the Jacobian matrix of f at (x_0, y_0) is non-degenerate, i.e., if $f = u + iv$, then

$$J_f(x_0,y_0) := \begin{pmatrix} \dfrac{\partial u}{\partial x}(x_0,y_0) & \dfrac{\partial u}{\partial y}(x_0,y_0) \\ \dfrac{\partial v}{\partial x}(x_0,y_0) & \dfrac{\partial v}{\partial y}(x_0,y_0) \end{pmatrix}$$

has non-zero determinant;
(3) for any pair of paths γ, $\tilde{\gamma}$ in U with $\gamma(t_0) = \tilde{\gamma}(\tilde{t}_0) = z_0$ that are smooth at t_0, $\tilde{t}_0$, respectively, the oriented angle between γ and $\tilde{\gamma}$ at z_0 relative to t_0, $\tilde{t}_0$ is equal to the oriented angle between $\Gamma := f \circ \gamma$ and $\tilde{\Gamma} := f \circ \tilde{\gamma}$ at $f(z_0)$ relative to t_0, $\tilde{t}_0$.

Remark 3.1. Notice that, owing to parts (1) and (2) of Definition 3.3, the paths Γ and $\tilde{\Gamma}$ are smooth at t_0, $\tilde{t}_0$, respectively, hence part (3) makes sense. Indeed, by the Chain Rule we have

$$\Gamma'(t_0) = \left(\frac{\partial u}{\partial x}(x_0, y_0)\gamma_1'(t_0) + \frac{\partial u}{\partial y}(x_0, y_0)\gamma_2'(t_0), \frac{\partial v}{\partial x}(x_0, y_0)\gamma_1'(t_0) + \frac{\partial v}{\partial y}(x_0, y_0)\gamma_2'(t_0) \right)$$
$$= \gamma'(t_0)J_f(x_0, y_0)^T \neq 0,$$

$$\tilde{\Gamma}'(\tilde{t}_0) = \left(\frac{\partial u}{\partial x}(x_0, y_0)\tilde{\gamma}_1'(\tilde{t}_0) + \frac{\partial u}{\partial y}(x_0, y_0)\tilde{\gamma}_2'(\tilde{t}_0), \frac{\partial v}{\partial x}(x_0, y_0)\tilde{\gamma}_1'(\tilde{t}_0) + \frac{\partial v}{\partial y}(x_0, y_0)\tilde{\gamma}_2'(\tilde{t}_0) \right)$$
$$= \tilde{\gamma}'(\tilde{t}_0)J_f(x_0, y_0)^T \neq 0,$$

where the superscript T denotes the transposed matrix (recall that $\gamma'(t_0)$ and $\tilde{\gamma}'(\tilde{t}_0)$ are row-vectors). The above calculation in fact shows that part (3) of Definition 3.3 is equivalent to saying that the matrix $J_f(x_0, y_0)$ preserves oriented angles between vectors (see Definition 3.2). It is also clear that if f is conformal at z_0 and continuously differentiable on U, then f^{-1} (which is defined and continuously differentiable on a neighbourhood of $f(z_0)$ by the Inverse Function Theorem) is conformal at $f(z_0)$.

If f is $\mathbb{C}$-differentiable at z_0 and $f'(z_0) \neq 0$, then $J_f(x_0, y_0)$ is the composition of the dilation by $|f'(z_0)|$ and the anti-clockwise rotation by $\arg(f'(z_0))$, and it is obvious that such a linear map preserves oriented angles between vectors. Thus, the following holds:

Proposition 3.1. *If f is $\mathbb{C}$-differentiable at z_0 and $f'(z_0) \neq 0$, then f is conformal at z_0.*

We will now extend Definition 3.3 to points in $\overline{\mathbb{C}}$. First, consider the map

$$\mathrm{rec} : \mathbb{C} \setminus \{0\} \to \mathbb{C} \setminus \{0\}, \quad z \mapsto \frac{1}{z},$$

which by Proposition 3.1 is conformal at every point of $\mathbb{C} \setminus \{0\}$ (explain!). Also, consider the composition $\mathrm{rec}_\Pi := \Pi^{-1} \circ \mathrm{rec} \circ \Pi$ from the set $S^2 \setminus \{N, 0\}$ onto itself, where Π is the stereographic projection.

Proposition 3.2. *The map rec_Π extends to the rotation of the sphere S^2 by the angle π around the axis $\{y = 0, w = 1/2\}$.*

Proof. Homework. $\square$

In what follows we will write rec_Π for the extended rotation map supplied by Proposition 3.2.

Definition 3.4. Let $z_0 \in \overline{\mathbb{C}}$ and let f be a map defined on a neighbourhood U of z_0 with values in $\overline{\mathbb{C}}$.

(1) If $z_0 \neq \infty$ and $f(z_0) \neq \infty$, then f is said to be conformal at z_0 if one can choose U with $\infty \notin U$, $\infty \notin f(U)$, such that the composition

$$f_1 := \Pi \circ f \circ \Pi^{-1} : \Pi(U) \to \mathbb{C}$$

is conformal at $\Pi(z_0)$ (here $f_1(\Pi(z_0)) = \Pi(f(z_0))$);

(2) if $z_0 \neq \infty$ and $f(z_0) = \infty$, then f is said to be conformal at z_0 if one can choose U with $\infty \notin U$, $\infty \notin \mathrm{rec}_\Pi(f(U))$, such that the composition

$$f_2 := \Pi \circ \mathrm{rec}_\Pi \circ f \circ \Pi^{-1} : \Pi(U) \to \mathbb{C}$$

is conformal at $\Pi(z_0)$ (here $f_2(\Pi(z_0)) = 0$);

(3) if $z_0 = \infty$ and $f(z_0) \neq \infty$, then f is said to be conformal at z_0 if one can choose U with $\infty \notin \mathrm{rec}_\Pi(U)$, $\infty \notin f(U)$, such that the composition

$$f_3 := \Pi \circ f \circ \mathrm{rec}_\Pi \circ \Pi^{-1} : \Pi(\mathrm{rec}_\Pi(U)) \to \mathbb{C}$$

is conformal at $\Pi(\mathrm{rec}_\Pi(z_0)) = 0$ (here $f_3(0) = \Pi(f(z_0))$);

(4) if $z_0 = \infty$ and $f(z_0) = \infty$, then f is said to be conformal at z_0 if one can choose U with $\infty \notin \mathrm{rec}_\Pi(U)$, $\infty \notin \mathrm{rec}_\Pi(f(U))$, such that the composition

$$f_4 := \Pi \circ \mathrm{rec}_\Pi \circ f \circ \mathrm{rec}_\Pi \circ \Pi^{-1} : \Pi(\mathrm{rec}_\Pi(U)) \to \mathbb{C}$$

is conformal at $\Pi(\mathrm{rec}_\Pi(z_0)) = 0$ (here $f_4(0) = 0$).

Remark 3.2. In Definition 3.4 we used the stereographic projection Π as well as the map rec_Π to reduce the situation to that of Definition 3.3. The use of rec_Π is justified by Proposition 3.2 since rotations of the sphere preserve any reasonably defined angles, and the use of Π can be justified too by accurately introducing oriented angles between paths in S^2 and showing that Π has the oriented angle preservation property. We do not explore these details here.

Definition 3.5. Let D, G be domains in $\overline{\mathbb{C}}$ and $f : D \to G$ a bijective map. Then f is said to be *a conformal map from D onto G*, or *a conformal map between D and G*, if f is conformal at every point of D. In this situation we also say that f *maps D conformally onto G*. If f is a conformal map from D onto G and f^{-1} is a conformal map from G onto D, then f is called *a conformal equivalence between D and G*. If there exists a conformal equivalence between D and G, the domains are said to be *conformally equivalent*. A conformal map from a domain D onto itself is called *a conformal transformation of D*. Finally, a map $g : D \to \overline{\mathbb{C}}$ is said to be *conformal on D* if $g(D)$ is a domain and g is a conformal map from D onto $g(D)$.

We will now state three important facts, which will be established later (see Lectures 10 and 17).

Theorem 3.1. *Any function holomorphic on a domain $D \subset \mathbb{C}$ is infinitely many times $\mathbb{C}$-differentiable at every point of D. In particular, functions holomorphic on D are continuously differentiable on D.*

Theorem 3.2. (The Open Mapping Theorem) *Let $D \subset \mathbb{C}$ be a domain and $f \in H(D)$. Assume that $f \not\equiv \mathrm{const}$. Then $f(D) \subset \mathbb{C}$ is a domain.*

Theorem 3.3. *Let $D \subset \mathbb{C}$ be a domain and $f \in H(D)$. If f is 1-to-1, then $f'(z) \neq 0$ for all $z \in D$.*

Using Proposition 3.1, Theorems 3.1, 3.2, 3.3 and the Inverse Function Theorem, one obtains:

Corollary 3.1. *Let $D \subset \mathbb{C}$ be a domain and $f \in H(D)$. If f is 1-to-1, then f is conformal on D. Furthermore, f is a conformal equivalence between D and $f(D)$.*

Proof. Homework. □

Exercises

3.1. Find the set $\Pi(R)$, where R is the intersection of S^2 with the plane $\{x = w\}$ in $\mathbb{R}^3$.

3.2. Describe $\Pi^{-1}(\Delta(1,2))$ as a subset of $\mathbb{R}^3$.

3.3. Prove that the limit set of the function $e^{\frac{1}{z}}$ (regarded as a map from $\mathbb{C} \setminus \{0\}$ to $\overline{\mathbb{C}}$) at 0 is all of $\overline{\mathbb{C}}$.

3.4. Let

$$\gamma(t) := \sqrt{2}e^{2\pi i t^3}, \quad \tilde{\gamma}(t) := \left(\sin t + 1 - \frac{1}{\sqrt{2}}\right)(1+i)$$

be two paths. Find the oriented angle between γ and $\tilde{\gamma}$ at the point $z_0 := (1+i)$ relative to $t_0 := 1/2$ and $\tilde{t}_0 := \pi/4$.

3.5. Determine which of the following maps from $\mathbb{C}$ to $\overline{\mathbb{C}}$ are conformal at 0:

(i) $z \mapsto \bar{z}$,

(ii) $z \mapsto z^2$,

(iii) $z \mapsto \begin{cases} \dfrac{z-1}{z} & \text{if } z \neq 0, \\ \infty & \text{if } z = 0, \end{cases}$

(iv) $z \mapsto z^3\bar{z}$,

(v) $z \mapsto ze^z$,

(vi) $z \mapsto \begin{cases} \dfrac{1}{z} & \text{if } z \neq 0, \\ \infty & \text{if } z = 0. \end{cases}$

3.6. Determine which of the following maps from $\overline{\mathbb{C}}$ to $\overline{\mathbb{C}}$ are conformal at ∞:

$$\text{(i) } z \mapsto \begin{cases} \bar{z} & \text{if } z \neq \infty, \\ \infty & \text{if } z = \infty, \end{cases}$$

$$\text{(ii) } z \mapsto \begin{cases} z^2 & \text{if } z \neq \infty, \\ \infty & \text{if } z = \infty, \end{cases}$$

$$\text{(iii) } z \mapsto \begin{cases} \dfrac{z-1}{z} & \text{if } z \neq 0, \infty, \\ \infty & \text{if } z = 0, \\ 1 & \text{if } z = \infty, \end{cases}$$

$$\text{(iv) } z \mapsto \begin{cases} z^3 \bar{z} & \text{if } z \neq \infty, \\ \infty & \text{if } z = \infty, \end{cases}$$

$$\text{(v) } z \mapsto \begin{cases} z e^z & \text{if } z \neq \infty, \\ \infty & \text{if } z = \infty, \end{cases}$$

$$\text{(vi) } z \mapsto \begin{cases} \dfrac{1}{\bar{z}} & \text{if } z \neq 0, \infty, \\ \infty & \text{if } z = 0 \\ 0 & \text{if } z = \infty. \end{cases}$$

3.7. Let $D \subset \mathbb{C}$ be a domain and $f \in H(D)$. Prove that $f' \in H(D)$.

3.8. Find all entire functions $f = u + iv$ such that $u = v^3 + 1$. (Hint: use Theorem 3.2.)

3.9. Let $D \subset \mathbb{C}$ be a domain and $f \in H(D)$. Assume that f' does not vanish at any point of D. Does it follow that f is 1-to-1? Prove your conclusion.

3.10. Using Corollary 3.1, show that the function $e^{\frac{1}{z}}$ is conformal on the domain

$$\left\{ z \in \mathbb{C} : \text{Im } z < 0, \ \left| z + \frac{i}{4\pi} \right| > \frac{1}{4\pi} \right\}.$$

Can you find the image of this domain under $e^{\frac{1}{z}}$?

Lecture 4
Conformal Maps (Continued). Möbius Transformations

We will now accumulate some examples of conformal maps between domains in $\mathbb{C}$ and, more generally, in $\overline{\mathbb{C}}$. Let us start with domains in $\mathbb{C}$ and first explore the exponential function $e^z = e^x(\cos y + i \sin y)$. As this function is entire, in order to see where it is conformal, by Corollary 3.1 we only need to understand on what domains in $\mathbb{C}$ it is 1-to-1.

Proposition 4.1. *We have $e^{z_1} = e^{z_2}$ if and only if $z_1 - z_2 = 2\pi i k$ for some $k \in \mathbb{Z}$.*

Proof. Write

$$e^{z_1} = e^{x_1}(\cos y_1 + i \sin y_1), \quad e^{z_2} = e^{x_2}(\cos y_2 + i \sin y_2).$$

Hence, $e^{z_1} = e^{z_2}$ if and only if

$$x_1 = x_2, \quad \cos y_1 = \cos y_2, \quad \sin y_1 = \sin y_2,$$

which occurs if and only if

$$x_1 = x_2, \quad y_1 - y_2 = 2\pi k \quad \text{for some } k \in \mathbb{Z}.$$

The proposition is proved. $\square$

By Proposition 4.1, the exponential function e^z is conformal on any horizontal strip $S_{\alpha,\beta} := \{z \in \mathbb{C} : \alpha < \operatorname{Im} z < \beta\}$ with $\beta - \alpha \leq 2\pi$. Consider an arbitrary horizontal line in $S_{\alpha,\beta}$, namely $\mathscr{L}_\delta := \{x + i\delta, x \in \mathbb{R}\}$, where $\alpha < \delta < \beta$. For every point $z \in \mathscr{L}_\delta$ we have

$$e^z = e^{x+i\delta} = e^x e^{i\delta},$$

with $e^{i\delta}$ being a point in the unit circle S^1. Therefore, $\mathscr{L}_\delta$ is mapped by e^z to the half-line of all points in $\mathbb{C}$ with argument $\delta \pmod{2\pi}$. Varying δ in the interval (α, β), we see that e^z maps $S_{\alpha,\beta}$ conformally onto the angle shown in Fig. 4.1 (here the strokes indicate *the complements* to the domains in question).

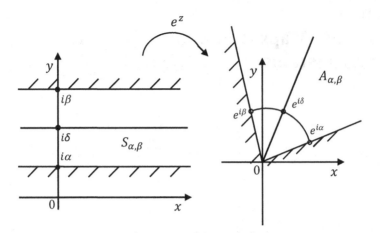

Fig. 4.1

We denote this angle by $A_{\alpha,\beta}$. It is the set of points located between the two half-lines emanating from the origin and passing through the points $e^{i\alpha}$, $e^{i\beta}$ as pictured in Fig. 4.1. In particular, the strip $S_{0,2\pi}$ is conformally mapped onto the angle $A_{0,2\pi} = \mathbb{C} \setminus \mathbb{R}_+$. We can therefore consider the conformal map inverse to the restriction of e^z to $S_{0,2\pi}$. It is natural to call this inverse a logarithm of some kind, and for $z \in \mathbb{C} \setminus \mathbb{R}_+$, we denote it by $\ln_0 z$.

Recall that in Lecture 1 we introduced $\ln z$ for any $z \neq 0$ as $\ln z = \ln|z| + i \arg z$ and showed that

$$e^{\ln z} = z \ \forall z \neq 0. \tag{4.1}$$

We will now prove that $\ln_0$ is just the restriction of $\ln$ to $A_{0,2\pi}$.

Proposition 4.2. *We have* $\ln_0 = \ln|_{A_{0,2\pi}}$.

Proof. We only need to show that $f(z) := \ln z|_{A_{0,2\pi}}$ is the inverse to $g(z) := e^z|_{S_{0,2\pi}}$. In view of identity (4.1) one has $g(f(z)) = z$ for all $z \in A_{0,2\pi}$, so it suffices to see that $f(g(z)) = z$ for every $z \in S_{0,2\pi}$. Write

$$f(g(z)) = \ln|e^z| + i \arg(e^z) = \ln e^x + iy = x + iy = z \ \forall z = x + iy \in S_{0,2\pi},$$

where in the second equality we utilised the fact that for all $z \in S_{0,2\pi}$ one has $\arg(e^z) = y$. □

Next, we consider the function z^n for $n = 2, 3, \ldots$. Again, this function is entire, so to find domains in $\mathbb{C}$ on which it is conformal, we only need to determine those where it is 1-to-1. Clearly, z^n is 1-to-1 on any angle $A_{\alpha,\beta}$, with $\beta - \alpha \leq 2\pi/n$, mapped by z^n onto the angle $A_{n\alpha,n\beta}$. In particular, $A_{0,2\pi/n}$ is conformally mapped by z^n onto the familiar angle $A_{0,2\pi} = \mathbb{C} \setminus \mathbb{R}_+$, and we can therefore look at the conformal map inverse to the restriction of z^n to $A_{0,2\pi/n}$. It is natural to call this inverse a root of order n of some kind, and for $z \in \mathbb{C} \setminus \mathbb{R}_+$, we denote it by $\sqrt[n]{z}$.

An explicit formula for $\sqrt[n]{z}$ is given below (cf. formula (1.2)).

Proposition 4.3. *We have* $\sqrt[n]{z} = |z|^{\frac{1}{n}} e^{i\frac{\arg z}{n}}$.

Proof. We need to show that $f(z) := |z|^{\frac{1}{n}} e^{i\frac{\arg z}{n}}|_{A_{0,2\pi}}$ is the inverse to $g(z) := z^n|_{A_{0,2\pi/n}}$. Let us start with proving that $g(f(z)) = z$ for all $z \in A_{0,2\pi}$. Write

$$g(f(z)) = \left(|z|^{\frac{1}{n}} e^{i\frac{\arg z}{n}}\right)^n = |z| e^{i\arg z} = z \ \forall z \in A_{0,2\pi}$$

as required. Notice that this part of the proof works for any $z \neq 0$.
Next,

$$f(g(z)) = f(|z^n| e^{i\arg z^n}) = |z| e^{i\frac{\arg z^n}{n}} = |z| e^{i\arg z} = z \ \forall z \in A_{0,2\pi/n},$$

where in the third equality we used the fact that for any element $z \in A_{0,2\pi/n}$ one has $\arg z^n = n \arg z$. □

In the above examples, $\ln_0 z$ and $\sqrt[n]{z}$ are the inverses to certain holomorphic 1-to-1 functions (restrictions of e^z and z^n) between domains in $\mathbb{C}$, hence they are conformal by Corollary 3.1. The natural question is then whether $\ln_0 z$ and $\sqrt[n]{z}$ are *holomorphic* on the respective domains. The answer is positive, as can be established using the explicit formulas for $\ln_0 z$ and $\sqrt[n]{z}$ from Propositions 4.2, 4.3. We will now obtain a general fact.

Theorem 4.1. *Let f be a $\mathbb{C}$-valued function defined on a neighbourhood of a point $z_0 \in \mathbb{C}$. Assume that f is conformal at z_0. Then f is $\mathbb{C}$-differentiable at z_0. Hence, if $D \subset \mathbb{C}$ is a domain and $g : D \to \mathbb{C}$ is conformal at every point of D, then $g \in H(D)$.*

Proof. As $f = u + iv$ is $\mathbb{R}$-differentiable at $z_0 = x_0 + iy_0$, we only need to verify the CR-equations

$$\frac{\partial u}{\partial x}(x_0, y_0) = \frac{\partial v}{\partial y}(x_0, y_0), \quad \frac{\partial u}{\partial y}(x_0, y_0) = -\frac{\partial v}{\partial x}(x_0, y_0).$$

Consider the Jacobian matrix

$$J_f(x_0, y_0) = \begin{pmatrix} \dfrac{\partial u}{\partial x}(x_0, y_0) & \dfrac{\partial u}{\partial y}(x_0, y_0) \\ \dfrac{\partial v}{\partial x}(x_0, y_0) & \dfrac{\partial v}{\partial y}(x_0, y_0) \end{pmatrix}$$

and recall that $\det J_f(x_0, y_0) \neq 0$. Then the CR-equations are equivalent to saying that the matrix $J_f(x_0, y_0)$ has the form

$$J_f(x_0, y_0) = C \begin{pmatrix} \cos \psi & -\sin \psi \\ \sin \psi & \cos \psi \end{pmatrix}$$

for some $C > 0$ and $0 \leq \psi < 2\pi$, which is the composition of a dilation and a rotation.

Taking into account that part (3) of Definition 3.3 is equivalent to the preservation of oriented angles between vectors by the matrix $J_f(x_0, y_0)$ (see Remark 3.1), we then conclude that the theorem is a consequence of the lemma stated below. □

Lemma 4.1. *Let A be a non-degenerate linear transformation of $\mathbb{R}^2$ preserving oriented angles between vectors. Then A is the composition of a dilation and a rotation.*

Proof. First, we will prove that the preservation of angles implies that A maps circles with centre 0 to circles with centre 0. Indeed, fix two vectors $v_1, v_2 \in \mathbb{R}^2$ of equal lengths not lying on the same line. Clearly, the oriented angle between v_1 and $v_1 + v_2$ is equal to that between $v_1 + v_2$ and v_2. Suppose that Av_1 and Av_2 have different lengths. Then one immediately observes that the oriented angle between Av_1 and $A(v_1 + v_2)$ differs from that between $A(v_1 + v_2)$ and Av_2 (see Fig. 4.2).

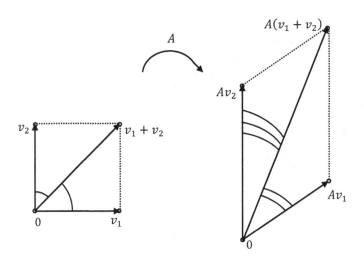

Fig. 4.2

This contradiction shows that the lengths of Av_1 and Av_2 are in fact equal, hence circles centred at 0 are mapped to circles centred at 0 as stated. Notice that so far we have not used the preservation of *oriented* angles.

Consider a circle

$$x^2 + y^2 = r^2$$

and write A^{-1} in matrix form as $A^{-1} = \begin{pmatrix} a & b \\ c & d \end{pmatrix}$, $a, b, c, d \in \mathbb{R}$. Then the image of the circle under A has the equation

$$(ax + by)^2 + (cx + dy)^2 = r^2,$$

or

$$(a^2 + c^2)x^2 + (b^2 + d^2)y^2 + 2(ab + cd)xy = r^2.$$

This last equation defines a circle if and only if

$$a^2 + c^2 = b^2 + d^2, \quad ab + cd = 0. \tag{4.2}$$

By (4.2) we have $d = \pm a$, $c = \mp b$. The top sign is what we need, so we must exclude the bottom one. Indeed, for the bottom sign the transformation A does not preserve the oriented angle between the standard basis vectors $e_1 = (1,0)$ and $e_2 = (0,1)$ (check!). This proves the lemma. $\square$

Remark 4.1. Let $D \subset \mathbb{C}$ be a domain and $f \in H(D)$ be 1-to-1. By Corollary 3.1 we know that f^{-1} is a conformal map from the domain $f(D)$ onto D, and Theorem 4.1 therefore implies $f^{-1} \in H(f(D))$. In fact, the $\mathbb{C}$-differentiability of f^{-1} at every point of $f(D)$ also follows by a slightly different argument not involving Theorem 4.1. Indeed, combining Theorems 3.1, 3.2, 3.3 with the Inverse Function Theorem, we see that f^{-1} is $\mathbb{R}$-differentiable (moreover, continuously differentiable) on the domain $f(D)$, with

$$J_{f^{-1}}(f(x,y)) = J_f(x,y)^{-1} \ \forall z = x + iy \in D.$$

Since $J_f(x,y)$ is the composition of a dilation and a rotation, the same holds for its inverse, i.e., for $J_{f^{-1}}(f(x,y))$. This shows that the CR-equations hold for f^{-1} on $f(D)$. In particular, one obtains that the functions $\ln_0 z$ and $\sqrt[n]{z}$ are holomorphic on $\mathbb{C} \setminus \mathbb{R}_+$ without using the explicit expressions from Propositions 4.2, 4.3.

Remark 4.2. We can now compute the derivatives of $\ln_0$ and $f(z) := \sqrt[n]{z}$. Indeed, by the Chain Rule one has

$$\ln_0(e^z) = z \ \forall z \in S_{0,2\pi} \Rightarrow \ln_0'(e^z)e^z = 1 \ \forall z \in S_{0,2\pi} \Rightarrow \ln_0'(w) = \frac{1}{w} \ \forall w \in A_{0,2\pi},$$

$$\sqrt[n]{z^n} = z \ \forall z \in A_{0,2\pi/n} \Rightarrow f'(z^n)nz^{n-1} = 1 \ \forall z \in A_{0,2\pi/n} \Rightarrow f'(w) = \frac{\sqrt[n]{w}}{nw} \ \forall w \in A_{0,2\pi}.$$

By Corollary 3.1 and Theorem 4.1 we have the following characterisation of the maps conformal on domains in $\mathbb{C}$:

Theorem 4.2. *Let D be a domain in $\mathbb{C}$. Then $f : D \to \mathbb{C}$ is a conformal map on D if and only if f is 1-to-1 and $f \in H(D)$. In this case $f^{-1} \in H(f(D))$ and f is a conformal equivalence between D and $f(D)$.*

We will now explore examples of conformal maps between domains in $\overline{\mathbb{C}}$ containing the point ∞. Before proceeding, we note that the last statement of Theorem 4.2 remains true for conformal maps between domains in $\overline{\mathbb{C}}$.

Proposition 4.4. *Let D, G be domains in $\overline{\mathbb{C}}$ and $f : D \to G$ a conformal map from D onto G. Then f is a conformal equivalence between D and G.*

Proof. Homework. □

Definition 4.1. *A Möbius transformation* is a function of the form

$$\lambda(z) = \frac{az+b}{cz+d}, \tag{4.3}$$

where $a, b, c, d \in \mathbb{C}$ satisfy

$$\det \begin{pmatrix} a & b \\ c & d \end{pmatrix} \neq 0. \tag{4.4}$$

If $c = 0$, then λ is just an affine function from $\mathbb{C}$ onto itself. Let us turn this affine function into a map of $\overline{\mathbb{C}}$ by means of Π. Namely, consider the composition

$$\lambda_\Pi := \Pi^{-1} \circ \lambda \circ \Pi.$$

It extends to a continuous map from $\overline{\mathbb{C}}$ onto itself, with $\lambda_\Pi(\infty) = \infty$ (check!). If, on the other hand, $c \neq 0$, then λ is a function from $\mathbb{C} \setminus \{-d/c\}$ onto $\mathbb{C} \setminus \{a/c\}$. Again, considering λ_Π instead of λ, we can extend it to a continuous map from $\overline{\mathbb{C}}$ onto itself, with $\lambda_\Pi(\Pi^{-1}(-d/c)) = \infty$, $\lambda_\Pi(\infty) = \Pi^{-1}(a/c)$ (check!). In the future, we will not distinguish between either λ, or λ_Π, or the extension of λ_Π to all of $\overline{\mathbb{C}}$ and simply write $\lambda(\infty) = \infty$ if $c = 0$ as well as $\lambda(-d/c) = \infty$, $\lambda(\infty) = a/c$ otherwise.

Next, assuming that $c \neq 0$, in formula (4.3) we divide the numerator by the denominator with remainder:

$$\lambda(z) = \frac{az+b}{cz+d} = s + \frac{r}{cz+d},$$

where $r \neq 0$ due to condition (4.4). This shows that λ is the composition of the following maps:

$$z \mapsto cz, \quad z \mapsto z+d, \quad z \mapsto \frac{1}{z} \text{ (the map rec)}, \quad z \mapsto rz, \quad z \mapsto z+s.$$

Each of them is a conformal map from $\overline{\mathbb{C}}$ onto itself (check!), thus we have proved:

Proposition 4.5. *Any Möbius transformation is a conformal map from $\overline{\mathbb{C}}$ onto itself.*

Exercises

4.1. Find a conformal map from the strip $\left\{ z \in \mathbb{C} : -\frac{\pi}{2} < \operatorname{Re} z < \frac{\pi}{2} \right\}$ onto the upper half-plane $H := \{ z \in \mathbb{C} : \operatorname{Im} z > 0 \}$.

4.2. Find a conformal map from the domain $H \setminus \{ z \in \mathbb{C} : \operatorname{Re} z = 0, \, 0 \leq \operatorname{Im} z \leq 1 \}$ onto H.

4.3. Find the images of the following curves under the map $z \mapsto z^2$:

$\quad$ (i) $\{z \in \mathbb{C} : \arg z = \varphi_0\},\ 0 \le \varphi_0 < 2\pi,$

$\quad$ (ii) $\{z \in \mathbb{C} : |z| = r_0\}, \quad r_0 > 0,$

$\quad$ (iii) $\{z \in \mathbb{C} : \operatorname{Re} z = a_0\},\ a_0 \in \mathbb{R},$

$\quad$ (iv) $\{z \in \mathbb{C} : \operatorname{Im} z = a_0\},\ a_0 \in \mathbb{R}.$

4.4. Find the images of the following domains under the map $z \mapsto z^2$:

$\quad$ (i) $\{z \in \mathbb{C} : \operatorname{Re} z > 0\},$

$\quad$ (ii) $\left\{z \in \mathbb{C} : \pi < \arg z < \dfrac{3\pi}{2}\right\},$

$\quad$ (iii) $\left\{z \in \mathbb{C} : \dfrac{5\pi}{4} < \arg z < \dfrac{3\pi}{2}\right\},$

$\quad$ (iv) $\{z \in \mathbb{C} : \operatorname{Im} z < -1\},$

$\quad$ (v) $\left\{z \in \mathbb{C} : |z| < 2,\, 0 < \arg z < \dfrac{\pi}{2}\right\},$

$\quad$ (vi) $\left\{z \in \mathbb{C} : |z| > \dfrac{1}{2},\, \operatorname{Re} z > 0\right\}.$

4.5. Prove that $\ln_0 z$ and $\sqrt[n]{z}$ are holomorphic on $\mathbb{C} \setminus \mathbb{R}_+$ and find their derivatives directly from the explicit formulas obtained in Propositions 4.2, 4.3.

4.6. Find a Möbius transformation λ satisfying

$$\lambda(i) = 0, \quad \lambda(\infty) = 1, \quad \lambda(-i) = \infty.$$

4.7. Prove that every Möbius transformation λ has at least one fixed point in $\overline{\mathbb{C}}$, and, unless λ is the identity map, no more than two fixed points.

4.8. Find the image of the disk $\Delta(1,2)$ under the following Möbius transformations:

$$\text{(i) } z \mapsto -2iz + 1,$$

$$\text{(ii) } z \mapsto \frac{2iz}{z+3},$$

$$\text{(iii) } z \mapsto \frac{z+1}{z-2},$$

$$\text{(iv) } z \mapsto \frac{z-1}{2z-6}.$$

4.9. Find the image of the half-pane $\{z \in \mathbb{C} : \operatorname{Re} z < 1\}$ under the following Möbius transformations:

$$\text{(i) } z \mapsto (1+i)z + 1,$$

$$\text{(ii) } z \mapsto \frac{z}{z-1+i},$$

(iii) $z \mapsto \dfrac{z}{z-2}$,

(iv) $z \mapsto \dfrac{4z}{z+1}$,

(v) $z \mapsto \dfrac{z-3+i}{z+1+i}$.

4.10. Using Proposition 3.1, show directly from Definition 3.4 that every Möbius transformation is conformal at every point of $\overline{\mathbb{C}}$.

Lecture 5
Möbius Transformations (Continued).
Generalised Circles. Symmetry

Notice that every Möbius transformation

$$\lambda(z) = \frac{az+b}{cz+d}$$

is given by the non-degenerate matrix $\begin{pmatrix} a & b \\ c & d \end{pmatrix}$, which is defined uniquely up to a complex non-zero multiple. We will now see that the composition of two Möbius transformations is the Möbius transformation given by the product of the matrices.

Proposition 5.1. *For any Möbius transformations*

$$\lambda(z) = \frac{az+b}{cz+d}, \quad \tilde{\lambda}(z) = \frac{\tilde{a}z+\tilde{b}}{\tilde{c}z+\tilde{d}}$$

the composition $\Lambda := \lambda \circ \tilde{\lambda}$ *is a Möbius transformation and is defined by the matrix* $\begin{pmatrix} a & b \\ c & d \end{pmatrix} \cdot \begin{pmatrix} \tilde{a} & \tilde{b} \\ \tilde{c} & \tilde{d} \end{pmatrix}$. *Hence, the inverse* λ^{-1} *is a Möbius transformation and is defined by the matrix* $\begin{pmatrix} a & b \\ c & d \end{pmatrix}^{-1}$, *or, equivalently, by the matrix* $\begin{pmatrix} d & -b \\ -c & a \end{pmatrix}$.

Proof. First, we compute the composition

$$\Lambda(z) = \frac{a\dfrac{\tilde{a}z+\tilde{b}}{\tilde{c}z+\tilde{d}}+b}{c\dfrac{\tilde{a}z+\tilde{b}}{\tilde{c}z+\tilde{d}}+d} = \frac{(a\tilde{a}+b\tilde{c})z+(a\tilde{b}+b\tilde{d})}{(c\tilde{a}+d\tilde{c})z+(c\tilde{b}+d\tilde{d})}.$$

On the other hand, we have

$$\begin{pmatrix} a & b \\ c & d \end{pmatrix} \cdot \begin{pmatrix} \tilde{a} & \tilde{b} \\ \tilde{c} & \tilde{d} \end{pmatrix} = \begin{pmatrix} a\tilde{a}+b\tilde{c} & a\tilde{b}+b\tilde{d} \\ c\tilde{a}+d\tilde{c} & c\tilde{b}+d\tilde{d} \end{pmatrix},$$

and the proof is complete.   □

Corollary 5.1. *Möbius transformations form a group with respect to composition, which is isomorphic to* $\mathrm{PGL}_2(\mathbb{C}) := \mathrm{GL}_2(\mathbb{C}) \Big/ \Big\{ \begin{pmatrix} a & 0 \\ 0 & a \end{pmatrix}, a \in \mathbb{C} \setminus \{0\} \Big\}.$

We will now study the important symmetry-preserving property of Möbius transformations.

Definition 5.1. *A* generalised circle *is either a circle of positive radius or a line in* $\mathbb{C}$.

Proposition 5.2. *Under the stereographic projection, there is a 1-to-1 correspondence between generalised circles and circles in* S^2, *i.e., the intersections of* S^2 *with planes in* $\mathbb{R}^3$ *passing through at least two points of* S^2. *If such a plane* $\mathscr{P}$ *does not pass through N, then* $\Pi(\mathscr{P} \cap S^2)$ *is a circle. Otherwise,* $\Pi((\mathscr{P} \cap S^2) \setminus \{N\})$ *is a line.*

Proof. Homework.   □

Owing to Proposition 5.2, generalised circles are often called *circles in* $\overline{\mathbb{C}}$. In what follows, we shall understand generalised circles either in the sense of Definition 5.1 or in the sense of Proposition 5.2 without explicitly distinguishing the two descriptions.

We will now show:

Proposition 5.3. *Any Möbius transformation maps every generalised circle to a generalised circle.*

Proof. First, consider the equation of a line $ax + by + c = 0$, with $a, b, c \in \mathbb{R}$, $a^2 + b^2 \neq 0$, and write it as

$$\frac{1}{2}(a - ib)z + \frac{1}{2}(a + ib)\bar{z} + c = 0,$$

or as

$$Az + \bar{A}\bar{z} + c = 0,$$

where $A \in \mathbb{C} \setminus \{0\}$. Further, the equation of a circle of positive radius r is $(x - a)^2 + (y - b)^2 = r^2$, which can be rewritten as

$$|z|^2 + Az + \bar{A}\bar{z} + c = 0,$$

with $A \in \mathbb{C}$, $c \in \mathbb{R}$, $c < |A|^2$. It then follows that we can express the equation of any generalised circle as

$$d|z|^2 + Az + \bar{A}\bar{z} + c = 0, \tag{5.1}$$

where $A \in \mathbb{C}$, $c, d \in \mathbb{R}$, $cd < |A|^2$. Conversely, every such equation defines a generalised circle (check!).

Now, it is clear from the argument in the last paragraph of Lecture 4, that the proposition only needs to be proved for $\lambda = \mathrm{rec}$ (i.e., for $\lambda(z) = 1/z$). Let C be a generalised circle written in the form (5.1). If $d \neq 0$ (i.e., C is a circle), then rec maps C into either a circle (if $c \neq 0$) or a line (if $c = 0$), where in the latter case $0 \in C$ is mapped to $\infty \in \lambda(C)$. If $d = 0$ (i.e., C is a line), then rec maps C into either a circle (if $c \neq 0$) or a line (if $c = 0$), where in the former case $\infty \in C$ is mapped to $0 \in \lambda(C)$ and in the latter case the points $0, \infty \in C \cap \lambda(C)$ are interchanged. □

We will now discuss symmetry with respect to generalised circles. We know what a pair of points symmetric with respect to a line is, so it remains to introduce an analogue of this concept for circles.

Definition 5.2. Let C be a circle of radius $r > 0$ centred at z_0. Two points $z, z' \in \mathbb{C}$ are said to be *symmetric with respect to* C, or z is said to be *symmetric to* z' *with respect to* C, or z' is said to be *symmetric to* z *with respect to* C, if

(1) $|z - z_0||z' - z_0| = r^2$ (in particular, $z \neq z_0$ and $z' \neq z_0$);
(2) z and z' both lie on a half-line coming out of the point z_0.

In the situation of Definition 5.2, to include pairs of points in $\overline{\mathbb{C}}$, we also say that z_0 and ∞ are symmetric with respect to C. In addition, we say that ∞ is symmetric to itself with respect to any line.

We will now give a geometric characterisation of symmetry with respect to a generalised circle.

Proposition 5.4. *Let C be a circle of positive radius centred at z_0. Then two points $z, z' \in \mathbb{C}$, with $z \neq z_0$, $z' \neq z_0$, are symmetric with respect to C if and only if they are the common points of all generalised circles passing through z and orthogonal to C. Analogously, if C is a line, then two points $z, z' \in \mathbb{C}$ are symmetric with respect to C if and only if they are the common points of all generalised circles passing through z and orthogonal to C.*

Proof. We only need to prove the proposition for the case when C is a circle. Fix any circle C' orthogonal to C and passing through z, and let z'' be the other point where C' intersects the half-line coming out of the point z_0 and containing z (see Fig. 5.1). Notice that if $z \in C$, one has $z'' = z$.

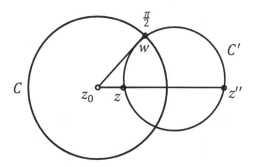

Fig. 5.1

Then the proposition is a consequence of the lemma stated below since it yields $|z - z_0||z'' - z_0| = |w - z_0|^2$. □

Lemma 5.1. *For any circle of positive radius, a point $A \in \mathbb{C}$ not contained in the open disk bounded by the circle, a tangent segment AB, and points C and D as shown in Fig. 5.2, we have $|AC||AD| = |AB|^2$.*

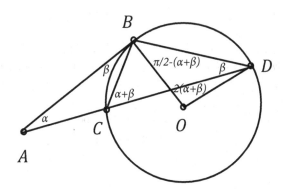

Fig. 5.2

Proof. The lemma trivially holds if A is a point of the circle, so we assume that A is not contained in the closed disk bounded by it.

It suffices to prove that the triangles $\triangle ABC$ and $\triangle ABD$ are similar since this implies

$$\frac{|AB|}{|AC|} = \frac{|AD|}{|AB|}.$$

Let O be the centre of the circle and $\alpha := \angle BAC$, $\beta := \angle ABC$. We clearly have $\angle BCD = \alpha + \beta$, and therefore $\angle BOD = 2(\alpha + \beta)$ (as the central angle supported by the same arc). We then see that $\angle OBD = \frac{\pi}{2} - (\alpha + \beta)$, which yields

$$\angle ABD = \frac{\pi}{2} + \angle OBD = \pi - (\alpha + \beta),$$

thus $\angle BDA = \beta$. This proves that $\triangle ABC$ and $\triangle ABD$ are similar. □

If one thinks of generalised circles as circles in S^2 (see Proposition 5.2), the statement of Proposition 5.4 can be extended to include pairs of symmetric points in all of $\overline{\mathbb{C}}$. Indeed, if C is a circle of positive radius with centre z_0, then every generalised circle passing through z_0 and orthogonal to C is a line. If we think about lines as circles in S^2 passing through ∞, the other point where all generalised circles passing through z_0 and orthogonal to C intersect is ∞, which agrees with Definition 5.2. If C is a line in $\mathbb{C}$, it is clear that all circles in S^2 passing through ∞ and orthogonal to C have only one common point, thus ∞ is symmetric to itself, which again agrees with

Definition 5.2. With this remark, Proposition 5.4 becomes a complete characterisation of symmetry, i.e., one does not in fact need to require that the points z, z' lie in
$\mathbb{C}$ and that neither of them coincides with z_0 in the case when C is a circle.

We have:

Corollary 5.2. *If C is a generalised circle and $z, z' \in \overline{\mathbb{C}}$ are symmetric with respect
to C, then for any Möbius transformation λ the points $\lambda(z)$, $\lambda(z')$ are symmetric
with respect to the generalised circle $\lambda(C)$.*

Proof. Homework. □

We will now utilise Corollary 5.2 to find all Möbius transformations of the unit
disk Δ, i.e., the Möbius transformations that map Δ onto itself.

Proposition 5.5. *All Möbius transformations of Δ are given by the formula*

$$\lambda(z) = e^{i\alpha} \frac{z - a}{1 - \bar{a}z}, \quad a \in \Delta, \; \alpha \in \mathbb{R}. \tag{5.2}$$

Proof. Let λ be a Möbius transformation of Δ and $a := \lambda^{-1}(0)$. It maps the point
$a' \in \overline{\mathbb{C}}$ symmetric to a with respect to $\partial \Delta = S^1$ to ∞. If $a = 0$, then $a' = \infty$, hence λ
is an affine transformation and therefore is just a rotation $\lambda(z) = e^{i\alpha}z$, with $\alpha \in \mathbb{R}$
(check!). If $a \neq 0$, then $a' = 1/\bar{a}$ (check!), so

$$\lambda(z) = A \frac{z - a}{z - \dfrac{1}{\bar{a}}} \text{ for some } A \in \mathbb{C},$$

or

$$\lambda(z) = \tilde{A} \frac{z - a}{1 - \bar{a}z} \text{ for some } \tilde{A} \in \mathbb{C}.$$

Since $|\lambda(z)| = 1$ for $|z| = 1$, it follows that $|\tilde{A}| = 1$ (check!), so λ has the form (5.2).
On the other hand, any Möbius transformation of this form clearly maps Δ onto
itself (check!). □

Next, we determine all Möbius transformations of the upper half-plane H.

Proposition 5.6. *All Möbius transformations of H are given by the formula*

$$\lambda(z) = \frac{az + b}{cz + d}, \quad a, b, c, d \in \mathbb{R}, \; ad - bc > 0. \tag{5.3}$$

Proof. The Möbius transformation

$$\lambda_0(z) := \frac{z - i}{z + i}$$

maps H onto Δ (check!). Therefore, any Möbius transformation of H is a composi-
tion of the form $\lambda_0^{-1} \circ \lambda \circ \lambda_0$, where λ is as in (5.2). By Proposition 5.1, to compute

this composition (let us call it Λ), we need to multiply the respective matrices as follows:

$$\begin{pmatrix} i & i \\ -1 & 1 \end{pmatrix} \begin{pmatrix} e^{i\alpha} & -e^{i\alpha}a \\ -\bar{a} & 1 \end{pmatrix} \begin{pmatrix} 1 & -i \\ 1 & i \end{pmatrix} = \begin{pmatrix} ie^{i\alpha}(1-a)+i(1-\bar{a}) & e^{i\alpha}(1+a)-(1+\bar{a}) \\ -e^{i\alpha}(1-a)+(1-\bar{a}) & ie^{i\alpha}(1+a)+i(1+\bar{a}) \end{pmatrix}.$$

Scaling the latter matrix by $-ie^{-i\alpha/2}$, we obtain a matrix with real entries, hence

$$\Lambda(z) = \frac{az+b}{cz+d} \text{ for some } a, b, c, d \in \mathbb{R}.$$

Further, as Λ maps the point i into H, we have $\operatorname{Im}\Lambda(i) > 0$, or, equivalently, $ad - bc > 0$ as in formula (5.3). On the other hand, any Möbius transformation of the form (5.3) clearly maps H onto itself. $\square$

Corollary 5.3. *The group of Möbius transformations of each of Δ and H with respect to composition is isomorphic to* $\mathrm{PSL}_2(\mathbb{R}) := \mathrm{SL}_2(\mathbb{R}) \Big/ \Big\{ \pm \begin{pmatrix} 1 & 0 \\ 0 & 1 \end{pmatrix} \Big\}$.

Exercises

5.1. Consider the circle $C := \{z \in \mathbb{C} : |z - i| = 2\}$. Find the circle in S^2 that corresponds to C in the way described in Proposition 5.2.

5.2. Find the point symmetric to $1 + i$ with respect to the circle $\{z \in \mathbb{C} : |z - 1| = 2\}$.

5.3. Find the inverse to any Möbius transformation λ of Δ as expressed by formula (5.2) and write λ^{-1} in the form (5.2). Give a formula for the parameters a and $e^{i\alpha}$ of λ^{-1} in terms of those of λ.

5.4. Find the composition of any Möbius transformations λ and $\tilde{\lambda}$ of H as expressed by formula (5.3) and write $\lambda \circ \tilde{\lambda}$ in the form (5.3). Give a formula for the parameters a, b, c, d of $\lambda \circ \tilde{\lambda}$ in terms of those of λ and $\tilde{\lambda}$.

5.5. For any $z_0 \in \Delta$, find all Möbius transformations of Δ that fix z_0 (write them in the form (5.2)).

5.6. For any $z_0 \in H$, find all Möbius transformations of H that fix z_0 (write them in the form (5.3)).

5.7. Prove that the function

$$\tan z := \frac{\sin z}{\cos z}$$

maps the strip $\left\{ z \in \mathbb{C} : -\frac{\pi}{4} < \operatorname{Re} z < \frac{\pi}{4} \right\}$ conformally onto Δ. (Hint: think of the function $\tan z$ as the composition of some previously studied maps.)

5.8. Find a conformal map from the domain introduced in Exercise 3.10 onto Δ.

5.9. Let C_1 and C_2 be two circles of positive radii such that C_2 lies in the open disk bounded by C_1. Further, let $\omega_1, \ldots, \omega_n$ be circles that lie between C_1 and C_2 and have the following properties: (i) for $j < n$ the circle ω_j is tangent to C_1, C_2 and ω_{j+1}, and (ii) ω_n is tangent to C_1, C_2 and ω_1. Prove that the number n is independent of the choice of the circles $\{\omega_j\}$. (Hint: reduce to the case when C_1 and C_2 are concentric.)

5.10. Let

$$J(z) := \frac{1}{2}\left(z + \frac{1}{z}\right),$$

with $J(0) := \infty$ and $J(\infty) := \infty$. Find all points in $\overline{\mathbb{C}}$ where J is conformal. Give examples of maximal domains on which J is conformal. Find the image of the right half-plane $\{z \in \mathbb{C} : \operatorname{Re} z > 0\}$ under J.

Lecture 6
Domains Bounded by Pairs of Generalised Circles. Integration

So far, we have accumulated the following examples of conformal maps:

(1) z^n, $n = 2,3,\ldots$, conformal on any angle $A_{\alpha,\beta}$, with $\beta - \alpha \le 2\pi/n$;
(2) $\sqrt[n]{z}$, $n = 2,3,\ldots$, conformal on $\mathbb{C} \setminus \mathbb{R}_+$;
(3) e^z, conformal on any strip $S_{\alpha,\beta}$, with $\beta - \alpha \le 2\pi$;
(4) $\ln_0$, conformal on $\mathbb{C} \setminus \mathbb{R}_+$;
(5) Möbius transformations, conformal on $\overline{\mathbb{C}}$.

As many complex-analytic concepts do not change under conformal equivalence, it is sometimes convenient to map a domain in $\overline{\mathbb{C}}$ conformally onto a domain of simpler form and carry out the required complex-analytic considerations in an easier setting. For instance, the Riemann Mapping Theorem (to be proved later) states that almost every "simply-connected" domain in $\overline{\mathbb{C}}$ is conformally equivalent to Δ. However, constructing such a conformal equivalence explicitly for a particular domain may be difficult. There are a variety of techniques for special classes of domains. For example, the maps listed in (1)–(5) above are useful for conformally simplifying domains bounded by pairs of generalised circles. We will now look at this situation in more detail.

Suppose we would like to map the domain D shown in Fig. 6.1 conformally onto the upper half-plane H, where, as before, the strokes indicate the complement to the domain in question. Apply the Möbius transformation

$$\lambda_0(z) := \frac{z - z_1}{z - z_2}.$$

It maps each circle into a line, and D is mapped onto an angle of size α with vertex at the origin, i.e., onto $A_{\beta,\beta+\alpha}$ for some $\beta \ge 0$. By rotating $A_{\beta,\beta+\alpha}$, we can assume that $\beta = 0$.

Suppose first that α is rational with respect to π, i.e., $\alpha = \pi p/q$ for positive integers p, q. If $p > 1$, we apply the function $\sqrt[p]{z}$, which is conformal on the angle $A_{0,\alpha}$ and maps it onto $A_{0,\pi/q}$. Next, z^q maps $A_{0,\pi/q}$ conformally onto H as required. Notice that the conformal map from D onto H is represented as the composition of certain mappings listed under (1), (2), (5). In particular, for $p > 1$ we utilised the composition

$$(z \mapsto z^q) \circ (z \mapsto \sqrt[p]{z}) \tag{6.1}$$

to map $A_{0,\alpha}$ onto H. Composition (6.1) should not be confused with the composition

$$(z \mapsto \sqrt[p]{z}) \circ (z \mapsto z^q),$$

which may not even be defined on $A_{0,\alpha}$ (explain!). We may think of function (6.1) as $z^{q/p}$. Indeed, by Proposition 4.3 it agrees with formula (1.1).

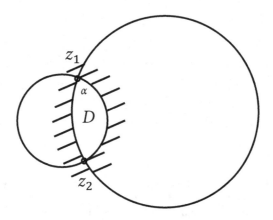

Fig. 6.1

Suppose that α is arbitrary. We then apply $\ln_0$ to $A_{0,\alpha}$ to obtain the strip $S_{0,\alpha}$, and the composition of the dilation $\pi z/\alpha$ and e^z clearly maps $S_{0,\alpha}$ onto H as required. Notice that the conformal map from D onto H in this case is represented as the composition of certain mappings listed under (3), (4), (5). In particular, we utilised the function $e^{\frac{\pi}{\alpha} \ln_0 z}$ to map $A_{0,\alpha}$ onto H. We may think of the above function as $z^{\pi/\alpha}$, which by Proposition 4.2 indeed agrees with formula (1.1). Thus, to conformally simplify domains bounded by pairs of generalised circles it is natural to use non-integral powers of z.

Let us look at a more specific example and see how the general approach outlined above works for it. Consider the domain D shown in Fig. 6.2. To find α, we first determine the radius, say r, of each of the circles, which clearly satisfies

$$1 + (r - (\sqrt{2} - 1))^2 = r^2$$

and therefore is equal to $\sqrt{2}$. It then follows that $\alpha = \pi/2$; in particular, α is rational with respect to π.

Now, the Möbius transformation

$$\lambda_0(z) := \frac{z-i}{z+i}$$

maps D onto the angle $A_{3\pi/4, 5\pi/4}$ as can be seen by computing

$$\lambda_0(\sqrt{2} - 1) = -\frac{1+i}{\sqrt{2}}, \quad \lambda_0(0) = -1$$

(check!). Next, the composition of the rotation $e^{-3\pi i/4}z$ and z^2 transforms $A_{3\pi/4, 5\pi/4}$ into H as required.

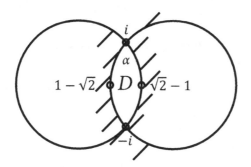

Fig. 6.2

We will now turn to integration of functions over paths in $\mathbb{C}$. For a subset $S \subset \mathbb{C}$ we denote by $C(S)$ the $\mathbb{C}$-vector space of all complex-valued functions continuous on S.

Definition 6.1. Let $D \subset \mathbb{C}$ be a domain. *A C^1-path in D is a C^1-map $\gamma : [0, 1] \to D$, where the derivatives of γ at 0 and 1 are understood as one-sided derivatives. The length of the C^1-path γ is defined as*

$$|\gamma| := \int_0^1 |\gamma'(t)| dt.$$

In fact, the definition of the length of a C^1-path given above agrees with the usual definition.

Proposition 6.1. *For a C^1-path γ one has*

$$|\gamma| = \sup \left\{ \sum_{j=1}^n |\gamma(t_j) - \gamma(t_{j-1})| : 0 =: t_0 < t_1 < \cdots < t_n := 1 \right\}, \quad (6.2)$$

where the supremum is taken over all partitions of $[0, 1]$.

Proof. Homework. (Hint: use the Mean Value Theorem for the functions $\gamma_1 = \operatorname{Re} \gamma$ and $\gamma_2 = \operatorname{Im} \gamma$.) $\square$

Definition 6.2. Let $D \subset \mathbb{C}$ be a domain, $f \in C(D)$ and $\gamma(t) = \gamma_1(t) + i\gamma_2(t)$ a C^1-path in D. *The integral of f over γ is*

$$\int_\gamma f(z)dz := \int_0^1 f(\gamma(t))\gamma'(t)dt = \int_0^1 f(\gamma_1(t) + i\gamma_2(t))(\gamma_1'(t) + i\gamma_2'(t))dt =$$

$$\int_0^1 \operatorname{Re}\left(f(\gamma_1(t) + i\gamma_2(t))(\gamma_1'(t) + i\gamma_2'(t))\right)dt +$$

$$i\int_0^1 \operatorname{Im}\left(f(\gamma_1(t) + i\gamma_2(t))(\gamma_1'(t) + i\gamma_2'(t))\right)dt.$$

We will often abbreviate $\int_\gamma f(z)dz$ as $\int_\gamma fdz$, and an analogous convention will be used for every type of integrals considered below.

Notice that the integral from Definition 6.2 is related to integrals of the kind $\int_\gamma Pdx + Qdy$ from real analysis. Indeed, if $f(z) = u(z) + iv(z)$, then

$$f(\gamma(t))\gamma'(t) = (u(\gamma(t)) + iv(\gamma(t)))(\gamma_1'(t) + i\gamma_2'(t)) =$$

$$(u(\gamma(t))\gamma_1'(t) - v(\gamma(t))\gamma_2'(t)) + i(v(\gamma(t))\gamma_1'(t) + u(\gamma(t))\gamma_2'(t)).$$

Hence

$$\int_\gamma fdz = \int_\gamma udx - vdy + i\int_\gamma vdx + udy. \tag{6.3}$$

In fact, one can allow P and Q to be complex-valued continuous functions (i.e., elements of $C(D)$) and make the integral $\int_\gamma Pdx + Qdy$ meaningful by separating the real and imaginary parts of the integrand. Then formula (6.3) turns into the natural formula

$$\int_\gamma fdz = \int_\gamma fdx + ifdy.$$

The following simple example of computing an integral will be of some importance later on:

Example 6.1. Let $D := \mathbb{C} \setminus \{0\}$, $f(z) := 1/z$, $\gamma(t) := e^{2\pi imt}$, with $m \in \mathbb{Z}$. Then one has $\gamma'(t) = 2\pi ime^{2\pi imt}$ (check!) and

$$\int_\gamma f(z)dz = \int_0^1 \frac{1}{e^{2\pi imt}} 2\pi ime^{2\pi imt}dt = 2\pi im.$$

We will now discuss some properties of integrals.

1. Linearity. We have

$$\int_\gamma (af + bg)dz = a\int_\gamma fdz + b\int_\gamma gdz$$

for all $f, g \in C(D)$ and $a, b \in \mathbb{C}$.

2. Independence of path re-parametrisation. Let $\psi : [0,1] \to [0,1]$ be a C^1-function with $\psi(0) = 0$, $\psi(1) = 1$ and let $\tilde{\gamma}(t) := \gamma(\psi(t))$. Then

$$\int_{\tilde{\gamma}} f dz = \int_\gamma f dz.$$

Indeed,

$$\int_{\tilde{\gamma}} f dz = \int_0^1 f(\gamma(\psi(t)))\gamma'(\psi(t))\psi'(t)dt = \int_0^1 f(\gamma(\tau))\gamma'(\tau)d\tau = \int_\gamma f dz.$$

Notice that the function ψ is *not* required to be monotone.

3. Dependence on the path orientation. Let $\gamma_-(t) := \gamma(1-t)$. The path γ_- is called *the reversal of γ*; we also say that γ_- is obtained from γ by *reversing the orientation of γ*. Then

$$\int_{\gamma_-} f dz = -\int_\gamma f dz.$$

The proof follows the same argument as in Property 2 with ψ replaced by $1-t$ (check!).

4. Summability over paths. Let $\tau_0 \in (0,1)$. Consider the paths $\tilde{\gamma}(t) := \gamma(t\tau_0)$ and $\tilde{\tilde{\gamma}}(t) := \gamma((1-t)\tau_0+t)$. Then

$$\int_\gamma f dz = \int_{\tilde{\gamma}} f dz + \int_{\tilde{\tilde{\gamma}}} f dz.$$

Indeed,

$$\int_\gamma f dz = \int_0^{\tau_0} f(\gamma(t))\gamma'(t)dt + \int_{\tau_0}^1 f(\gamma(t))\gamma'(t)dt =$$

$$\int_0^1 f(\gamma(t\tau_0))\tau_0\gamma'(t\tau_0)dt + \int_0^1 f(\gamma((1-t)\tau_0+t))(1-\tau_0)\gamma'((1-t)\tau_0+t)dt =$$

$$\int_0^1 f(\tilde{\gamma}(t))\tilde{\gamma}'(t)dt + \int_0^1 f(\tilde{\tilde{\gamma}}(t))\tilde{\tilde{\gamma}}'(t)dt = \int_{\tilde{\gamma}} f dz + \int_{\tilde{\tilde{\gamma}}} f dz.$$

5. An upper bound on the absolute value. One has

$$\left| \int_\gamma f dz \right| \le \int_0^1 |f(\gamma(t))||\gamma'(t)|dt. \tag{6.4}$$

In particular, if $|f(z)| \le M$ for all $z \in \gamma([0,1])$, then

$$\left| \int_\gamma f dz \right| \le M|\gamma|.$$

To obtain (6.4), let $I := \int_{\gamma} f\,dz$ and write $I = |I|e^{i\theta}$ for a suitable $\theta \in \mathbb{R}$. Then

$$|I| = e^{-i\theta}I = \int_{\gamma} e^{-i\theta} f(z)\,dz = \int_0^1 e^{-i\theta} f(\gamma(t))\gamma'(t)\,dt.$$

Since $|I| \in \mathbb{R}$, we in fact have

$$|I| = \int_0^1 \mathrm{Re}\left(e^{-i\theta} f(\gamma(t))\gamma'(t)\right)dt \le \int_0^1 \left|e^{-i\theta} f(\gamma(t))\gamma'(t)\right|dt = \int_0^1 |f(\gamma(t))||\gamma'(t)|dt$$

as required.

Exercises

6.1. Find a conformal map from the domain shown in Fig. 6.3 onto Δ.

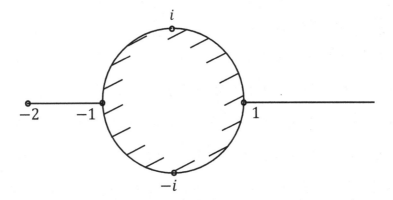

Fig. 6.3

Note that in Fig. 6.3 the closed disk $\overline{\Delta}$, the segment $[-2, -1]$ and the half-line $[1, \infty)$ are *deleted* from $\mathbb{C}$.

6.2. Find a conformal map from $\Delta \setminus [0, 1)$ onto Δ.

6.3. Which of the following paths are C^1-paths:

$$\text{(i) } e^{2\pi i f(t)},$$

$$\text{(ii) } t^2 + ig(t),$$

$$\text{(iii) } \sin^2 t + ih(t)\,?$$

Here

$$f(t) := \begin{cases} t^2 & \text{if } 0 \le t \le \dfrac{1}{2}, \\ \dfrac{t}{2} & \text{if } \dfrac{1}{2} \le t \le 1, \end{cases}$$

$$g(t) := \begin{cases} \dfrac{\sin t}{t} & \text{if } 0 < t \le 1, \\ 1 & \text{if } t = 0, \end{cases}$$

$$h(t) := \begin{cases} e^{-\frac{1}{(t-1/2)^2}} & \text{if } 0 \le t \le 1, t \ne \dfrac{1}{2}, \\ 0 & \text{if } t = \dfrac{1}{2}. \end{cases}$$

Prove your conclusions.

6.4. For $z_1, z_2 \in \mathbb{C}$ we think of the segment $[z_1, z_2]$ as the path $(1-t)z_1 + tz_2$. Prove that $[z_1, z_2]$ is a C^1-path and that $|[z_1, z_2]| = |z_1 - z_2|$.

6.5. Compute the integrals

$$\text{(i)} \int_\gamma (z - i)dz, \text{ where } \gamma(t) := t + it^2,$$

$$\text{(ii)} \int_\gamma \frac{1}{z^2} dz, \quad \text{where } \gamma(t) := e^{\pi i(t-1)}.$$

6.6. Using Property 4 of integrals, find

$$\int_\gamma |x + y| dz,$$

where $\gamma(t) := (1 + i)(2t - 1)$.

6.7. Let $D := \{z \in \mathbb{C} : 1/2 < |z| < 3/2\}$. For a function $f \in C(D)$ and a C^1-path γ in D consider the integral $\int_\gamma f dz$. Among the following paths in D, determine all pairs for which the corresponding integrals are equal for every function $f \in C(D)$ and all pairs for which the integrals may differ depending on the choice of such f:

$$\text{(i)} \ e^{2\pi i t},$$

$$\text{(ii)} \ e^{2\pi i t^2},$$

$$\text{(iii)} \ e^{2\pi i \sin\left(\frac{5}{2}\pi t\right)},$$

$$\text{(iv)} \ e^{2\pi i \cos\left(\frac{5}{2}\pi t\right)},$$

$$\text{(v)} \ e^{4\pi i t^3},$$

$$\text{(vi)} \ e^{4\pi i t^4}.$$

Prove your conclusions.

6.8. Let $P(z) = z^K + a_{K-1}z^{K-1} + \cdots + a_0$ be a polynomial in z with leading coefficient 1. Prove that

$$\max_{|z|=1} |P(z)| \geq 1.$$

(Hint: after dividing P by a certain quantity you may find Property 5 of integrals useful.)

6.9. Prove that there does not exist a polynomial $P(z) = \sum_{j=0}^{K} a_j z^j$ in z such that

$$\max_{|z|=1} \left| P(z) - \frac{1}{z^2} \right| < 1.$$

(Hint: the idea is similar to that for doing Exercise 6.8.)

6.10. Find a counterexample to the following statement: if $D \subset \mathbb{C}$ is a domain, $f \in C(D)$ and $\{\gamma^n\}$ a sequence of C^1-paths in D that converges to a C^1-path γ in D uniformly on $[0, 1]$, then

$$\lim_{n \to \infty} \int_{\gamma^n} f(z)dz = \int_{\gamma} f(z)dz.$$

6.11. Let f be $\mathbb{R}$-differentiable at 0. Prove that

$$\lim_{\varepsilon \to 0} \frac{1}{\varepsilon^2} \int_{|z|=\varepsilon} f(z)dz = 2\pi i \frac{\partial f}{\partial \bar{z}}(0),$$

where $|z| = \varepsilon$ is a shorthand for the path $\varepsilon e^{2\pi i t}$.

Lecture 7

Primitives Along Paths. Holomorphic Primitives. The Existence of a Holomorphic Primitive of a Function Holomorphic on a Disk. Goursat's Lemma

We will now look at methods for computing integrals.

Definition 7.1. Let $D \subset \mathbb{C}$ be a domain, $f \in C(D)$ and γ a C^1-path in D. A function $\Phi : [0, 1] \to \mathbb{C}$ is called *a primitive of f along γ* if Φ is differentiable everywhere on $[0, 1]$ and $\Phi'(t) = f(\gamma(t))\gamma'(t)$ for all $t \in [0, 1]$.

Clearly, if f has a primitive Φ along γ, it follows that

$$\int_\gamma f(z)dz = \Phi(1) - \Phi(0).$$

Definition 7.2. Let $D \subset \mathbb{C}$ be a domain and $f : D \to \mathbb{C}$. A function $F : D \to \mathbb{C}$ is called *a holomorphic primitive of f [on D]* if $F \in H(D)$ and $F'(z) = f(z)$ for all $z \in D$.

Notice that by Theorem 3.1 the existence of a holomorphic primitive for f implies, by differentiating the primitive twice, that $f \in H(D)$.

If F is a holomorphic primitive of f on D and γ is a C^1-path in D, then $\Phi(t) := F(\gamma(t))$ is a primitive of f along γ. Indeed, one has

$$\Phi'(t) \stackrel{\text{check!}}{=\!=\!=\!=} F'(\gamma(t))\gamma'(t) = f(\gamma(t))\gamma'(t) \ \forall t \in [0, 1].$$

Thus, the following holds:

Proposition 7.1. *Let $D \subset \mathbb{C}$ be a domain and assume that $f : D \to \mathbb{C}$ has a holomorphic primitive F on D. Then for every C^1-path γ in D we have*

$$\int_\gamma f(z)dz = F(\gamma(1)) - F(\gamma(0)).$$

In particular, in this case $\int_\gamma f dz$ depends only on the endpoints of γ; hence, if γ is a closed path (i.e., $\gamma(1) = \gamma(0)$), then $\int_\gamma f dz = 0$.

Even if f is holomorphic on a domain D, it may not have a holomorphic primitive there as can be seen, for instance, from Example 6.1. Indeed, if $f(z) = 1/z$

had a holomorphic primitive on $D = \mathbb{C} \setminus \{0\}$, then its integral over the closed path $\gamma(t) = e^{2\pi i m t}$ would be zero for all $m \in \mathbb{Z}$, which is not the case.

Nevertheless, for a function $f \in H(D)$ one can construct a primitive along any C^1-path γ in D by following the four steps outlined below.

Step 1. Partition the segment $[0,1]$ into subsegments as $0 = t_0 < t_1 < \cdots < t_n = 1$ so that for $j = 0, \ldots, n-1$ the set $\gamma([t_j, t_{j+1}])$ lies in an open disk $\Delta^j \subset D$. The existence of such a partition follows from the uniform continuity of γ as well as from the existence of $\delta > 0$ for which $\Delta(\gamma(t), \delta)$ lies in D for all $t \in [0,1]$. To see where δ comes from, for any two non-empty subsets $S_1, S_2 \subset \mathbb{C}$ let

$$\mathrm{dist}(S_1, S_2) := \inf_{z_1 \in S_1, z_2 \in S_2} |z_1 - z_2|,$$

and set $\mathrm{dist}(S, \emptyset) := \infty$ for all $S \subset \mathbb{C}$. Also, for $z \in \mathbb{C}$ and any subset $S \subset \mathbb{C}$ define $\mathrm{dist}(z, S) := \mathrm{dist}(\{z\}, S)$. We have a general fact:

Proposition 7.2. *Let $\mathbf{K} \subset \mathbb{C}$ be compact. Then for any $S \subset \mathbb{C}$ there exists $z_0 \in \mathbf{K}$ such that $\mathrm{dist}(\mathbf{K}, S) = \mathrm{dist}(z_0, S)$. Hence, if $\mathbf{K} \cap S = \emptyset$ and S is closed, one has $\mathrm{dist}(\mathbf{K}, S) > 0$.*

In other words, the distance between two subsets one of which is compact is attained at some point of the compact one.

Proof. Homework. (Hint: prove that the function $\rho(z) := \mathrm{dist}(z, S)$ is continuous on $\mathbb{C}$.) $\square$

By Proposition 7.2, $\mathrm{dist}(\gamma([0,1]), \partial D) \neq 0$, so we choose $0 < \delta \leq \mathrm{dist}(\gamma([0,1]), \partial D)$.

Step 2. Construct a holomorphic primitive, say F_j, of f on Δ^j, for $j = 0, \ldots, n-1$ (as will be shown later on, this is indeed possible).

Step 3. Match F_j and F_{j+1} on $\Delta^j \cap \Delta^{j+1}$ for $j = 0, \ldots, n-2$. Namely, for the function $\psi_j := F_{j+1} - F_j$ on $\Delta^j \cap \Delta^{j+1}$ we have $\psi_j' = F_{j+1}' - F_j' = f - f \equiv 0$, which implies $\psi_j \equiv C$, with $C \in \mathbb{C}$ (check!). Choosing a new primitive $\tilde{F}_{j+1}$ on Δ^{j+1} as $\tilde{F}_{j+1} := F_{j+1} - C$, we then see that $\tilde{F}_{j+1}$ matches F_j on $\Delta^j \cap \Delta^{j+1}$. Proceeding from $j = 0$ to $j = n-2$ yields a matching as required. Note that if $\Delta^k \cap \Delta^\ell \neq \emptyset$ for some $\ell > k+1$, it may happen that $F_k \neq F_\ell$ on $\Delta^k \cap \Delta^\ell$ despite the fact that $F_j = F_{j+1}$ on $\Delta^j \cap \Delta^{j+1}$ for $j = 0, \ldots, n-2$.

Step 4. Introduce a function Φ on $[0,1]$ as follows: if $t_j \leq t \leq t_{j+1}$ for some integer $0 \leq j \leq n-1$, set

$$\Phi(t) := F_j(\gamma(t)).$$

First of all, we need to check that this function is well-defined. Indeed, apart from the segment $[t_j, t_{j+1}]$, the point t_j also lies in $[t_{j-1}, t_j]$ (provided $j > 0$), so we must ascertain that $F_{j-1}(\gamma(t_j)) = F_j(\gamma(t_j))$. The latter, however, is ensured by the matching that we performed in Step 3. Similarly, apart from the segment $[t_j, t_{j+1}]$, the point t_{j+1} also lies in $[t_{j+1}, t_{j+2}]$ (provided $j < n-1$), and we again have $F_j(\gamma(t_{j+1})) = F_{j+1}(\gamma(t_{j+1}))$ by Step 3.

We will now show that Φ is a primitive of f along γ. Fix $\tau_0 \in [0,1]$ and find an integer $0 \leq j \leq n-1$ such that $t_j \leq \tau_0 \leq t_{j+1}$. Then on an interval containing the point τ_0 we have $\Phi(t) = F_j(\gamma(t))$. Therefore,

$$\Phi'(\tau_0) = F_j'(\gamma(\tau_0))\gamma'(\tau_0) = f(\gamma(\tau_0))\gamma'(\tau_0)$$

as required.

To illustrate the method for constructing primitives along paths outlined above, we shall now revisit Example 6.1 for $m = 1$.

Example 7.1. Let $f(z) := 1/z$, $D := \mathbb{C} \setminus \{0\}$, $\gamma(t) := e^{2\pi i t}$. Partition $[0,1]$ by the points $t_0 = 0, t_1 := 1/8, t_2 := 1/4, t_3 := 3/8, t_4 := 1/2, t_5 := 5/8, t_6 := 3/4, t_7 := 7/8,$ $t_8 = 1$ (here $n = 8$) and choose Δ^j as shown in Fig. 7.1, $j = 0, \ldots, 7$ (notice that Δ^j is centred at $\gamma(t_j)$ although this is not a requirement in Step 1).

Now, we fix holomorphic primitives F_j of f on each of the disks Δ^j. By Remark 4.2, we see that $\ln_0$ is a holomorphic primitive of f on $\mathbb{C} \setminus \mathbb{R}_+$. We also consider the function

$$\widetilde{\ln}_0 z := \ln|z| + i\,\widetilde{\arg}\,z$$

defined on $\mathbb{C} \setminus \mathbb{R}_-$, with $\mathbb{R}_- := \{x \in \mathbb{R} : x \leq 0\}$, where $\widetilde{\arg}$ is the analogue of $\arg$ measured from $-\pi$ to π rather than from 0 to 2π. Arguing as in the proof of Proposition 4.2, one can show that $\widetilde{\ln}_0$ is the inverse to $e^z|_{S_{-\pi,\pi}}$, and, as in Remark 4.2, we again see that $\widetilde{\ln}_0$ is a holomorphic primitive of f on $\mathbb{C} \setminus \mathbb{R}_-$. Notice that $\widetilde{\ln}_0 z = \ln_0 z$ if $\operatorname{Im} z > 0$ and $\widetilde{\ln}_0 z = \ln_0 z - 2\pi i$ if $\operatorname{Im} z < 0$.

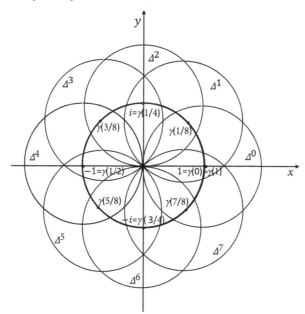

Fig. 7.1

Set $F_j := \widetilde{\ln_0}|_{\Delta^j}$ for $j = 0, 1, 2$, $F_j := \ln_0|_{\Delta^j}$ for $j = 3, 4, 5, 6$, $F_7 := \widetilde{\ln_0}|_{\Delta^7} + 2\pi i$. We can now construct a primitive Φ of f along γ as in Step 4. For this primitive we have $\Phi(0) = F_0(1) = \widetilde{\ln_0} 1 = 0$, $\Phi(1) = F_7(1) = \widetilde{\ln_0} 1 + 2\pi i = 2\pi i$. Therefore

$$\int_\gamma f(z)dz = \Phi(1) - \Phi(0) = 2\pi i,$$

which agrees with the calculation in Example 6.1 for $m = 1$. The above construction can be easily modified to accommodate any value of m by introducing multiple copies of the disks Δ^j (provide full details!).

To justify Step 2 of the procedure for building primitives along paths, we shall now prove:

Theorem 7.1. *A function holomorphic on an open disk has a holomorphic primitive on the disk.*

Proof. Without loss of generality we can assume that the disk in question is Δ. For the proof, we will consider closed triangles contained in Δ. If T is such a triangle, its boundary ∂T is the union of three segments. We assign an orientation to each segment so that the overall orientation of ∂T is anti-clockwise (see Fig. 7.2). Each of the oriented segments will be regarded as a path with the usual parametrisation; e.g., we think of $[z_1, z_2]$ as the path $(1-t)z_1 + tz_2$. Then the ordinary length of each segment coincides with its length as a C^1-path from Definition 6.1 (see Exercise 6.4), and the sum of the lengths is exactly the perimeter of T, which we denote by $|\partial T|$ (below this notation will be applied to any triangle in $\mathbb{C}$).

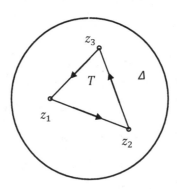

Fig. 7.2

Now, for $f \in C(\Delta)$ define *the integral of f over the boundary ∂T of T* as

$$\int_{\partial T} f(z)dz := \int_{[z_1,z_2]} fdz + \int_{[z_2,z_3]} fdz + \int_{[z_3,z_1]} fdz. \tag{7.1}$$

Later in the course this definition will be also utilised for triangles in an arbitrary domain $D \subset \mathbb{C}$ and functions continuous on D.

We need the following fact:

Lemma 7.1. *Let $f \in C(\Delta)$ and assume that for any triangle T contained in Δ one has $\int_{\partial T} f \, dz = 0$. Then f has a holomorphic primitive on Δ.*

Proof. For $z \in \Delta$ set

$$F(z) := \int_{[0,z]} f(\zeta) d\zeta.$$

We will now show that $F \in H(\Delta)$ and $F' = f$ on Δ. Fix $z_0 \in \Delta$ and for all sufficiently small Δz calculate

$$\frac{F(z_0 + \Delta z) - F(z_0)}{\Delta z} - f(z_0) = \frac{1}{\Delta z} \left(\int_{[0,z_0+\Delta z]} f(\zeta) d\zeta - \int_{[0,z_0]} f(\zeta) d\zeta \right) - f(z_0) =$$

$$\frac{1}{\Delta z} \left(\int_{[0,z_0+\Delta z]} f(\zeta) d\zeta + \int_{[z_0,0]} f(\zeta) d\zeta \right) - f(z_0) = \frac{1}{\Delta z} \int_{[z_0,z_0+\Delta z]} f(\zeta) d\zeta - f(z_0),$$

where the last equality follows by considering the triangle T with vertices at 0, z_0, $z_0 + \Delta z$ and using the condition $\int_{\partial T} f \, dz = 0$ (check!).

We now write the number $f(z_0)$ as

$$f(z_0) = \frac{1}{\Delta z} \int_{[z_0,z_0+\Delta z]} f(z_0) d\zeta,$$

hence

$$\frac{1}{\Delta z} \int_{[z_0,z_0+\Delta z]} f(\zeta) d\zeta - f(z_0) = \frac{1}{\Delta z} \int_{[z_0,z_0+\Delta z]} (f(\zeta) - f(z_0)) d\zeta.$$

Therefore

$$\left| \frac{F(z_0 + \Delta z) - F(z_0)}{\Delta z} - f(z_0) \right| = \frac{1}{|\Delta z|} \left| \int_{[z_0,z_0+\Delta z]} (f(\zeta) - f(z_0)) d\zeta \right| \leq$$

$$\frac{1}{|\Delta z|} \max_{\zeta \in [z_0,z_0+\Delta z]} |f(\zeta) - f(z_0)| \, |[z_0, z_0 + \Delta z]| = \max_{\zeta \in [z_0,z_0+\Delta z]} |f(\zeta) - f(z_0)|.$$

By the continuity of f we have

$$\max_{\zeta \in [z_0,z_0+\Delta z]} |f(\zeta) - f(z_0)| \to 0 \text{ as } \Delta z \to 0,$$

and the lemma follows. $\square$

Now, Theorem 7.1 is a consequence of Lemma 7.1 and the lemma stated below. $\square$

Lemma 7.2. *(Goursat's Lemma) Let $D \subset \mathbb{C}$ be a domain and $f \in H(D)$. Then for any triangle $T \subset D$ one has $\int_{\partial T} f \, dz = 0$.*

We will prove this lemma in the next lecture.

Exercises

7.1. Let
$$\Phi(t) := \frac{11 - 2i}{3} t^3 - (3 + 4i)t^2 + (-1 + 2i)t, \ t \in [0, 1].$$

Find a function for which Φ is a primitive along the segment $[1, 2i]$.

7.2. Find holomorphic primitives of the following functions on the respective domains:

(i) e^z on $\mathbb{C}$,

(ii) $\sin z$ on $\mathbb{C}$,

(iii) $\cos z$ on $\mathbb{C}$,

(iv) $\dfrac{1}{\cos^2 z}$ on $\mathbb{C} \setminus \left\{ \dfrac{\pi}{2} + \pi k, k \in \mathbb{Z} \right\}$.

7.3. Compute the integrals from Exercise 6.5 using holomorphic primitives.

7.4. Which of the following functions have holomorphic primitives on $\mathbb{C} \setminus \{0\}$:

(i) $\sin \dfrac{1}{z}$,

(ii) $\cos \dfrac{1}{z}$,

(iii) $e^{\frac{1}{z}}$,

(iv) $e^{\frac{1}{z^3}}$?

Prove your conclusions. (Hint: expand each of the functions into a series in powers of z and notice that of all the powers only $1/z$ has a problem with the existence of a holomorphic primitive on $\mathbb{C} \setminus \{0\}$.)

7.5. Let $D \subset \mathbb{C}$ be a domain not containing 0. Assume that the function $1/z$ has a holomorphic primitive, say F, on D. Prove that there exists $c \in \mathbb{C}$ such that
$$F(z) + c \in \text{Ln } z \ \forall z \in D.$$

7.6. Construct a primitive of the function $e^{\frac{1}{z}}$ along the path $\gamma(t) := e^{4\pi i t}$ and use it to compute the integral
$$\int_\gamma e^{\frac{1}{z}} dz.$$

7.7. Let $g : [0, 1] \to \mathbb{C}$ be a differentiable function, and consider the function f introduced in Exercise 2.23 by formula (2.3). Prove that for any $\tau_0 \in (0, 1)$ there exist the limits

$$A := \lim_{z \to \tau_0, \, \mathrm{Im}\, z > 0} f(z),$$
$$B := \lim_{z \to \tau_0, \, \mathrm{Im}\, z < 0} f(z),$$

and $A - B = g(\tau_0)$. (Hint: for t near τ_0 write $g(t) = g(\tau_0) + g'(\tau_0)(t - \tau_0) + o(t - \tau_0)$ and notice that only the first term matters; to analyse this term you may wish to use logarithms as in Example 7.1.)

7.8. Let T be the triangle with vertices $i, -1, 1 - i$. Find the integrals over ∂T of the following functions:

$$\text{(i) } \bar{z},$$

$$\text{(ii) } \frac{1}{z},$$

$$\text{(iii) } z^2,$$

$$\text{(iv) } z^3 \bar{z}^2,$$

$$\text{(v) } \frac{1}{\bar{z}^2},$$

$$\text{(vi) } e^{z^2},$$

$$\text{(vii) } \overline{f(\bar{z})},$$

where f is any entire function.

7.9. Let $D \subset \mathbb{C}$ be a domain, $f \in C(D)$, and T a triangle in D. Suppose that T is divided into finitely many subtriangles $T_1, \ldots, T_n$ with non-intersecting interiors. Prove that

$$\int_{\partial T} f \, dz = \sum_{j=1}^{n} \int_{\partial T_j} f \, dz.$$

7.10. Let $f \in C(\Delta)$ and define

$$F(z) := \int_{[0,z]} f(\zeta) \, d\zeta, \quad z \in \dot{\Delta}.$$

Prove that F is $\mathbb{C}$-differentiable at 0 and find $F'(0)$.

Lecture 8

Proof of Lemma 7.2. Constructible Primitives of Holomorphic Functions along Paths. Integration of Holomorphic Functions over Arbitrary Paths. Homotopy. Simply-Connected Domains. The Riemann Mapping Theorem

We will now prove Goursat's Lemma stated at the end of Lecture 7.

Proof (Lemma 7.2). Fix a triangle $T \subset D$. Using the mid-points of the sides, we split T into four subtriangles T', T'', T''', T'''' (see Fig. 8.1).

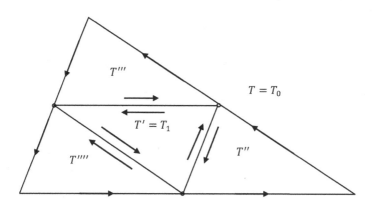

Fig. 8.1

Recalling Exercise 7.9, we have

$$\int_{\partial T} f\,dz = \int_{\partial T'} f\,dz + \int_{\partial T''} f\,dz + \int_{\partial T'''} f\,dz + \int_{\partial T''''} f\,dz$$

(notice that the integrals over the newly introduced segments cancel out due to the fact that each such segment occurs in two copies with opposite orientations).

Let $M := |\int_{\partial T} f\,dz|$. Then for at least one of the subtriangles T', T'', T''', T'''' the absolute value of the integral of f over the boundary is greater than or equal to $M/4$.

Suppose that this is the case for T'. We then set $T_0 := T$, $T_1 := T'$ and partition T_1 into four subtriangles as we have done above for T_0 (see Fig. 8.2).

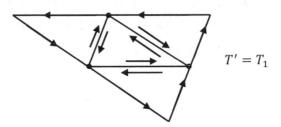

$T' = T_1$

Fig. 8.2

Among the four subtriangles of T_1 there is at least one for which the absolute value of the integral over the boundary is greater than or equal to $M/16$. Choose one subtriangle with the above property and call it T_2. Continuing this process, we construct a sequence $\{T_n\}$ of triangles with $\operatorname{diam}(T_n) \to 0$ as $n \to \infty$ and such that for $n = 0, 1, \dots$ we have:

$$(1) \ T_{n+1} \subset T_n;$$

$$(2) \ |\partial T_n| = \frac{|\partial T_0|}{2^n};$$

$$(3) \ \left| \int_{\partial T_n} f(z)dz \right| \geq \frac{M}{4^n}.$$

In particular, $\{T_n\}$ is a sequence of nested compact subsets of $\mathbb{C}$ whose diameters tend to zero, and by Cantor's Intersection Theorem we see that the intersection $\cap_{n=0} T_n$ is non-empty and in fact consists of a single point. Let z_0 be this point and write f as

$$f(z) = f(z_0) + f'(z_0)(z - z_0) + o(z - z_0).$$

Then

$$\int_{\partial T_n} f(z)dz = \int_{\partial T_n} f(z_0)dz + \int_{\partial T_n} f'(z_0)(z - z_0)dz + \int_{\partial T_n} o(z - z_0)dz.$$

The functions $f(z_0)$ and $f'(z_0)(z - z_0)$ have holomorphic primitives on all $\mathbb{C}$ (check!), hence, using Proposition 7.1 we see that the first two integrals are equal to zero (check!). Next, write $o(z - z_0)$ as

$$o(z - z_0) = \alpha(z)(z - z_0),$$

where the function $\alpha(z)$ tends to 0 as $z \to z_0$. Then we have

$$\frac{M}{4^n} \le \left| \int_{\partial T_n} f(z)dz \right| = \left| \int_{\partial T_n} \alpha(z)(z-z_0)dz \right| \le \max_{z \in \partial T_n} |\alpha(z)(z-z_0)| |\partial T_n| \le$$

$$\max_{z \in \partial T_n} |\alpha(z)| \max_{z \in \partial T_n} |z - z_0| |\partial T_n| \le \max_{z \in \partial T_n} |\alpha(z)| |\partial T_n|^2 = \max_{z \in \partial T_n} |\alpha(z)| \frac{|\partial T_0|^2}{4^n}.$$

Therefore, for $n = 0, 1, 2, \ldots$ one obtains

$$M \le \max_{z \in \partial T_n} |\alpha(z)| |\partial T_0|^2.$$

Recalling that ∂T_n accumulates to z_0 as $n \to \infty$ and that $\alpha(z) \to 0$ as $z \to z_0$, we see that the right-hand side in the above inequality tends to zero as $n \to \infty$. Hence $M = 0$, and the lemma is proved. □

Now that we have established Theorem 7.1, we arrive at:

Corollary 8.1. *If $D \subset \mathbb{C}$ is a domain, $f \in H(D)$ and γ is a C^1-path in D, then f has a primitive along γ.*

Notice that the four-step procedure for producing a primitive of a holomorphic function f along a path outlined in Lecture 7 can in fact be applied to *any* path γ (not necessarily a C^1-path), and in what follows any function Φ arising from this procedure will be called *a constructible primitive of f along γ*. We can now introduce the integral of a holomorphic function over any path.

Definition 8.1. Let $D \subset \mathbb{C}$ be a domain, $f \in H(D)$, γ any path in D and $\Phi : [0,1] \to \mathbb{C}$ a constructible primitive of f along γ. Define *the integral of f over γ* as

$$\int_\gamma f(z)dz := \Phi(1) - \Phi(0). \tag{8.1}$$

Note that a constructible primitive along a path is not unique (explain!). Therefore, to justify the definition we need:

Proposition 8.1. *The right-hand side of (8.1) is independent of the choice of Φ.*

Proof. Homework. (Hint: show that for any two choices of Φ their difference is locally constant, hence constant, on $[0,1]$ – see Exercise 8.2.) □

Remark 8.1. It is not hard to verify that Properties 1–4 of integrals stated in Lecture 6 remain valid for the integral introduced in Definition 8.1, where in Property 2 the function ψ is assumed to be monotone and continuous (check!).

There is the following analogue of Proposition 7.1:

Proposition 8.2. *Let $D \subset \mathbb{C}$ be a domain and suppose that $f : D \to \mathbb{C}$ has a holomorphic primitive F on D. Then for any path γ in D we have*

$$\int_\gamma f(z)dz = F(\gamma(1)) - F(\gamma(0)).$$

In particular, in this case $\int_\gamma f\,dz$ depends only on the endpoints of γ; hence, if γ is a closed path (i.e., $\gamma(1) = \gamma(0)$), then $\int_\gamma f\,dz = 0$.

Proof. Homework. □

Our next goal is to state and prove Cauchy's theorem on independence of integrals of holomorphic functions of homotopy. We start by defining homotopic paths. We will give two definitions: one for paths with common endpoints (*PCE-homotopy*), the other for closed paths (*CP-homotopy*).

Definition 8.2. (PCE-homotopy) Let $D \subset \overline{\mathbb{C}}$ be a domain and γ, $\tilde{\gamma}$ two paths in D with $\gamma(0) = \tilde{\gamma}(0) = z_0$, $\gamma(1) = \tilde{\gamma}(1) = z_1$ for some $z_0, z_1 \in D$. The path γ is said to be *PCE-homotopic [in D] to $\tilde{\gamma}$* if there exists a continuous map $\Gamma : [0,1] \times [0,1] \to D$, called *a PCE-homotopy between γ and $\tilde{\gamma}$ [in D]*, such that the following holds:

(1) $\Gamma(t,0) = \gamma(t)$, $\Gamma(t,1) = \tilde{\gamma}(t)$ $\forall t \in [0,1]$;
(2) $\Gamma(0,s) = z_0$, $\Gamma(1,s) = z_1$ $\forall s \in [0,1]$.

Observe that if $\Gamma(t,s)$ is a PCE-homotopy between γ and $\tilde{\gamma}$, then $\Gamma(t,1-s)$ is a PCE-homotopy between $\tilde{\gamma}$ and γ, so if γ is PCE-homotopic to $\tilde{\gamma}$, we see that $\tilde{\gamma}$ is PCE-homotopic to γ. This means that PCE-homotopy is a symmetric binary relation, and in the above situation we often say that *γ and $\tilde{\gamma}$ are PCE-homotopic [in D]* or that *γ and $\tilde{\gamma}$ are homotopic [in D] as paths with common endpoints.*

Remark 8.2. It is not hard to show that if γ is PCE-homotopic to $\tilde{\gamma}$ and $\tilde{\gamma}$ is PCE-homotopic to $\tilde{\tilde{\gamma}}$, then γ is PCE-homotopic to $\tilde{\tilde{\gamma}}$ (check!). Thus, PCE-homotopy is in fact an equivalence relation.

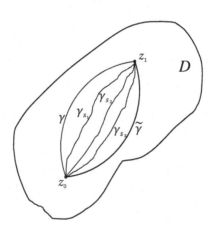

Fig. 8.3

Loosely speaking, the conditions of Definition 8.2 guarantee that one can "continuously deform" γ and $\tilde{\gamma}$ into one another within D by way of the paths

$\gamma_s(t) := \Gamma(t,s)$, which are required to have the same endpoints as γ and $\tilde{\gamma}$ (see Fig. 8.3).

Remark 8.3. Let γ be a path in a domain D and $\tilde{\gamma}(t) := \gamma(\psi(t))$, where $\psi : [0,1] \to [0,1]$ is a continuous bijection (hence a homeomorphism). Then we have either (i) $\psi(0) = 0$, $\psi(1) = 1$ and ψ is increasing or (ii) $\psi(0) = 1$, $\psi(1) = 0$ and ψ is decreasing. In the first case the paths γ and $\tilde{\gamma}$ are obtained from one another by re-parametrisation, have common endpoints, and $\Gamma(t,s) := \gamma((1-s)t + s\psi(t))$ is a PCE-homotopy between γ and $\tilde{\gamma}$ in D. In the second case, we consider the reversal $\tilde{\gamma}_-(t) = \tilde{\gamma}(1-t) = \gamma(\psi(1-t))$ and analogously conclude that γ and $\tilde{\gamma}_-$ are homotopic in D as paths with common endpoints. Note that the above arguments work for any continuous function $\psi : [0,1] \to [0,1]$ satisfying either (i) or (ii).

Definition 8.3. (CP-homotopy) Let $D \subset \overline{\mathbb{C}}$ be a domain and γ, $\tilde{\gamma}$ two paths in D that are closed (i.e., $\gamma(0) = \gamma(1)$, $\tilde{\gamma}(0) = \tilde{\gamma}(1)$). The path γ is said to be *CP-homotopic* [*in D*] *to* $\tilde{\gamma}$ if there exists a continuous map $\Gamma : [0,1] \times [0,1] \to D$, called *a CP-homotopy between γ and $\tilde{\gamma}$* [*in D*], such that the following holds:

(1) $\Gamma(t,0) = \gamma(t)$, $\Gamma(t,1) = \tilde{\gamma}(t)$ $\forall t \in [0,1]$;
(2) $\Gamma(0,s) = \Gamma(1,s)$ $\forall s \in [0,1]$.

If $\Gamma(t,s)$ is a CP-homotopy between γ and $\tilde{\gamma}$, then $\Gamma(t,1-s)$ is a CP-homotopy between $\tilde{\gamma}$ and γ, so if γ is CP-homotopic to $\tilde{\gamma}$, we see that $\tilde{\gamma}$ is CP-homotopic to γ. This means that CP-homotopy is a symmetric binary relation, and in the above situation we often say that *γ and $\tilde{\gamma}$ are CP-homotopic* [*in D*] or that *γ and $\tilde{\gamma}$ are homotopic* [*in D*] *as closed paths*.

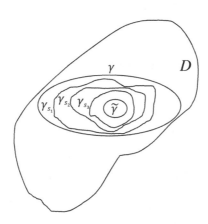

Fig. 8.4

Remark 8.4. It is easy to see that if γ is CP-homotopic to $\tilde{\gamma}$ and $\tilde{\gamma}$ is CP-homotopic to $\tilde{\tilde{\gamma}}$, then γ is CP-homotopic to $\tilde{\tilde{\gamma}}$ (check!). Thus, CP-homotopy is an equivalence relation.

The conditions of Definition 8.3 make sure that one can "continuously deform" γ and $\tilde{\gamma}$ into one another within D by way of the paths $\gamma_s(t) := \Gamma(t,s)$, which are required to be closed (see Fig. 8.4).

Remark 8.5. Let γ be a closed path in a domain D. It defines a continuous map $\hat{\gamma}$ from the unit circle S^1 to D as follows: $\hat{\gamma}(e^{2\pi it}) := \gamma(t)$ for all $t \in [0,1]$. Suppose that $\tilde{\gamma}$ is another closed path in D with $\hat{\tilde{\gamma}} = \hat{\gamma} \circ \Psi$, where $\Psi : S^1 \to S^1$ is a continuous bijection (hence a homeomorphism). One can show (check!) that there exists a continuous bijection $\psi : [0,1] \to [0,1]$ and $0 \le \alpha < 2\pi$ such that $\Psi(e^{2\pi it}) = e^{i\alpha} e^{2\pi i\psi(t)}$ for all $t \in [0,1]$, where we have either (i) $\psi(0) = 0$, $\psi(1) = 1$ and ψ is increasing or (ii) $\psi(0) = 1$, $\psi(1) = 0$ and ψ is decreasing. In the first case the paths γ and $\tilde{\gamma}$ are CP-homotopic. Indeed, for all $t \in [0,1]$ one has

$$\tilde{\gamma}(t) = \hat{\tilde{\gamma}}(e^{2\pi it}) = \hat{\gamma}(\Psi(e^{2\pi it})) = \hat{\gamma}\left(e^{i\alpha} e^{2\pi i\psi(t)}\right) = \gamma\left(\left(\psi(t) + \frac{\alpha}{2\pi}\right) (\mathrm{mod}\, 1)\right).$$

Set

$$\Gamma(t,s) := \gamma\left(\left(\psi(t) + (1-s)\frac{\alpha}{2\pi}\right) (\mathrm{mod}\, 1)\right).$$

This function is a CP-homotopy between $\tilde{\gamma}$ and γ_0 in D, where γ_0 is defined as $\gamma_0(t) := \gamma(\psi(t)(\mathrm{mod}\, 1)) = \gamma(\psi(t))$ (check!), and a CP-homotopy between γ_0 and γ is constructed as in Remark 8.3. These two CP-homotopies lead to a CP-homotopy between γ and $\tilde{\gamma}$. In the second case (i.e., when $\psi(0) = 1$, $\psi(1) = 0$), we consider the reversal

$$\tilde{\gamma}_-(t) = \tilde{\gamma}(1-t) = \gamma\left((\psi(1-t) + \frac{\alpha}{2\pi})(\mathrm{mod}\, 1)\right)$$

and analogously conclude that γ and $\tilde{\gamma}_-$ are homotopic in D as closed paths. Note that the above arguments work for any continuous function $\psi : [0,1] \to [0,1]$ satisfying either (i) or (ii).

Observe that if γ and $\tilde{\gamma}$ are both closed and have common endpoints, they may be homotopic in D as closed paths but not as paths with common endpoints (see Fig. 8.5).

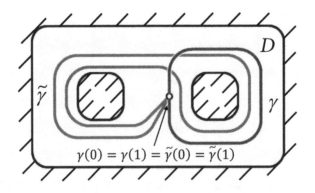

$$\gamma(0) = \gamma(1) = \tilde{\gamma}(0) = \tilde{\gamma}(1)$$

Fig. 8.5

When γ, $\tilde{\gamma}$ are either PCE-homotopic or CP-homotopic and there is no fear of confusion as to in which sense the homotopy is understood, we just say that γ *and* $\tilde{\gamma}$ *are homotopic* [*in D*], or that γ *is homotopic* [*in D*] *to* $\tilde{\gamma}$, or that $\tilde{\gamma}$ *is homotopic* [*in D*] *to* γ.

Proposition 8.3. *Let $D \subset \overline{\mathbb{C}}$ be a domain. Then the following two conditions are equivalent:*

(1) *Any two paths in D with common endpoints are PCE-homotopic in D;*

(2) *Any two closed paths in D are CP-homotopic in D, or, equivalently, every closed path in D is CP-homotopic in D to some (hence every) point in D regarded as a constant closed path.*

Proof. Homework. (Hint: to obtain the implication $(2) \Rightarrow (1)$, for paths γ and $\tilde{\gamma}$ with common endpoints consider the closed path constructed by "joining" γ and $\tilde{\gamma}_-$.) □

Definition 8.4. A domain $D \subset \overline{\mathbb{C}}$ is called *simply-connected* if it satisfies any of the two equivalent conditions of Proposition 8.3.

We will now state without proof a characterisation of simply-connected domains.

Theorem 8.1. *A domain $D \subset \overline{\mathbb{C}}$ is simply-connected if and only if any of the equivalent conditions given below is satisfied:*

(1) $\overline{\mathbb{C}} \setminus D$ *is connected;*

(2) *the boundary of D in $\overline{\mathbb{C}}$ is connected.*

Notice that if D is a domain in $\mathbb{C}$ then condition (1) of Theorem 8.1 cannot be replaced with the requirement that $\mathbb{C} \setminus D$ be connected and (2) cannot be replaced with the requirement that the boundary of D in $\mathbb{C}$ be connected. An elementary example for both statements is provided by the unbounded domain $\mathscr{D} := \{z \in \mathbb{C} : |z| > 1\}$, which is not simply-connected. Note that $\mathbb{C} \setminus \mathscr{D} = \overline{\Delta}$ is connected, whereas $\overline{\mathbb{C}} \setminus \mathscr{D} = \overline{\Delta} \cup \{\infty\}$ is disconnected. Similarly, the boundary of $\mathscr{D}$ in $\mathbb{C}$ is the connected set S^1, whereas the boundary of $\mathscr{D}$ in $\overline{\mathbb{C}}$ is $S^1 \cup \{\infty\}$, which is disconnected. However, for *bounded* domains in the complex plane Theorem 8.1 implies:

Theorem 8.2. *A bounded domain $D \subset \mathbb{C}$ is simply-connected if and only if any of the equivalent conditions stated below is satisfied:*

(1) $\mathbb{C} \setminus D$ *is connected;*

(2) *the boundary of D in $\mathbb{C}$ is connected.*

The importance of simply-connected domains is demonstrated, for instance, by the following major result, which will be obtained at the end of the course (see Lecture 21):

Theorem 8.3. (The Riemann Mapping Theorem) *Let $D \subset \overline{\mathbb{C}}$ be a simply-connected domain such that $\overline{\mathbb{C}} \setminus D$ contains at least two points. Then D is conformally equivalent to Δ.*

We mention without providing details that the sufficiency implications of Theorems 8.1, 8.2 may be deduced by an argument relying, in particular, on the proof of Theorem 8.3 (see Remark 21.1).

Exercises

8.1. Explain in detail what would have gone wrong in the proof of Lemma 7.2 had we not supposed that the function f is holomorphic on D. Give an example showing that the lemma may not hold if f is only assumed to be $\mathbb{R}$-differentiable on D. Also, give an example of a domain $D \subset \mathbb{C}$, a C^∞-function f on D, with $f \notin H(D)$, and a triangle $T \subset D$, such that $\int_{\partial T} f\,dz = 0$.

8.2. Let X be a connected topological space and Y a set. Assume that a mapping $f : X \to Y$ is locally constant, i.e., for every $x \in X$ there exist $y \in Y$ and an open subset $U \subset X$ containing x such that $f(x') = y$ for all $x' \in U$. Prove that f is constant, i.e., there exists $y_0 \in Y$ with $f(x) = y_0$ for all $x \in X$.

8.3. For any $z_0 \neq 0$ and $l \in \operatorname{Ln} z_0$, find a (not necessarily C^1-) path γ in $\mathbb{C} \setminus \{0\}$ joining 1 and z_0 such that

$$\int_\gamma \frac{dz}{z} = l.$$

(Hint: consider going around the origin a number of times plus going from 1 to z_0 along a line.)

8.4. Find the integral

$$\int_\gamma \sin \frac{2}{z}\,dz,$$

where

$$\gamma(t) := \begin{cases} e^{2\pi i t \sin \frac{\pi}{2t}} & \text{if } 0 < t \le 1, \\ 1 & \text{if } t = 0. \end{cases}$$

Is γ a C^1-path? Attempt to compute the value of the expression in the right-hand side of formula (6.2) for this path.

8.5. Let $D \subset \mathbb{C}$ be a domain and $f \in H(D)$. Prove that for every $z_0, z_1 \in D$, every path γ in D joining z_0 and z_1 (here $\gamma(0) = z_0$, $\gamma(1) = z_1$), and every $n \in \mathbb{N}$ the following holds:

$$f(z_1) = f(z_0) + \sum_{k=1}^{n} \frac{f^{(k)}(z_0)}{k!}(z_1 - z_0)^k + \frac{1}{n!}\int_\gamma (z_1 - \zeta)^n f^{(n+1)}(\zeta)d\zeta,$$

where $f^{(k)}$ is the kth derivative of f, which exists by Theorem 3.1 for all $k \in \mathbb{N}$. (Hint: use induction on n.)

8.6. Prove that the paths $\gamma(t) := e^{2\pi i t}$ and $\tilde{\gamma}(t) := 1 + 2e^{2\pi i t}$ are homotopic in $\mathbb{C} \setminus \{0\}$ (as closed paths).

8.7. Is the map

$$\Gamma(t,s) := (1-s)e^{\pi i(t-1/2)} + se^{-\pi i(t+1/2)}$$

a PCE-homotopy between $\gamma(t) := e^{\pi i(t-1/2)}$ and $\tilde{\gamma}(t) := e^{-\pi i(t+1/2)}$ in the following domains:

(i) $\mathbb{C}$,

(ii) $\mathbb{C} \setminus \{0\}$,

(iii) $\mathbb{C} \setminus \{1\}$,

(iv) $\mathbb{C} \setminus \overline{\Delta(3,1)}$?

Prove your conclusions.

8.8. Is the map

$$
\Gamma(t,s) := \begin{cases} \dfrac{\left(s-\dfrac{1}{2}\right)\sin 2\pi t}{\left(s-\dfrac{1}{2}\right)^2 + \sin^2 2\pi t}(1+i) & \text{if } t,s \in [0,1],\, s \neq \dfrac{1}{2}, \\[4mm] 0 & \text{if } t \in [0,1],\, s = \dfrac{1}{2} \end{cases}
$$

a PCE-homotopy between

$$
\gamma(t) := -\frac{2\sin 2\pi t}{1+4\sin^2 2\pi t}(1+i)
$$

and

$$
\tilde{\gamma}(t) := \frac{2\sin 2\pi t}{1+4\sin^2 2\pi t}(1+i)
$$

in $\mathbb{C}$? Prove your conclusion.

8.9. Is the map

$$
\Gamma(t,s) := 4 + (5-4s)e^{2\pi i t}
$$

a CP-homotopy between $\gamma(t) := 4 + 5e^{2\pi i t}$ and $\tilde{\gamma}(t) := 4 + e^{2\pi i t}$ in the following domains:

(i) $\mathbb{C}$,

(ii) $\mathbb{C} \setminus \{0\}$,

(iii) $\mathbb{C} \setminus \{9\}$,

(iv) $\mathbb{C} \setminus [10,15]$?

Prove your conclusions.

8.10. Construct a PCE-homotopy between $\gamma(t) := e^{2\pi i t}$ and $\tilde{\gamma}(t) := e^{2\pi i \sin\left(\frac{5\pi t}{2}\right)}$ in $\mathbb{C} \setminus \{0\}$.

8.11. Which of the following domains in $\overline{\mathbb{C}}$ are simply-connected:

(i) Δ_5,

(ii) $\{z \in \mathbb{C} : 1 < |z| < 2\}$,

(iii) $\{z \in \mathbb{C} : |z| > 1\} \cup \{\infty\}$,

(iv) $\{z \in \mathbb{C} : |z| > 1\} \setminus \overline{\Delta(2,1)}$,

(v) $\{z \in \mathbb{C} : |z| > 1\} \cup \{\infty\} \setminus \overline{\Delta(2,1)}$,

(vi) $\{z \in \mathbb{C} : |z| > 1\} \setminus \overline{\Delta(3,1)}$,

(vii) $\{z \in \mathbb{C} : |z| > 1\} \cup \{\infty\} \setminus \overline{\Delta(3,1)}$,

(viii) $\Delta_7 \setminus (\overline{\Delta(3,1)} \cup \overline{\Delta(5,1)})$?

Prove your conclusions.

8.12. A domain $D \subset \mathbb{C}$ is called *star-shaped* if there exists a point $z_0 \in D$ such that for every $z \in D$ the segment $[z_0, z]$ is contained in D. Show that every star-shaped domain is simply-connected.

8.13. Give an example of a topological space X and an open connected subset $U \subset X$ such that $X \setminus U$ is connected but ∂U is disconnected.

Lecture 9
Cauchy's Independence of Homotopy Theorem. Integration over Piecewise C^1-paths. Jordan Domains and Integration over their Boundaries

We will now state and prove one of the central theorems of the course.

Theorem 9.1. (Cauchy's Independence of Homotopy Theorem) *Let $D \subset \mathbb{C}$ be a domain and $f \in H(D)$. If γ and $\tilde{\gamma}$ are two paths in D that are homotopic in D either as paths with common endpoints or as closed paths, then $\int_\gamma f \, dz = \int_{\tilde{\gamma}} f \, dz$.*

Proof. By the definition of either kind of homotopy, there exists a continuous map $\Gamma : [0,1] \times [0,1] \to D$ such that $\Gamma(t,0) = \gamma(t)$ and $\Gamma(t,1) = \tilde{\gamma}(t)$ for all $t \in [0,1]$. Set $\gamma_s(t) := \Gamma(t,s)$ and $I(s) := \int_{\gamma_s} f \, dz$, $s \in [0,1]$. To prove the theorem, it suffices to show that the function $I(s)$ is locally constant on $[0,1]$, i.e., for every $s_0 \in [0,1]$ there is a neighbourhood of s_0 in $[0,1]$ on which $I(s)$ is constant. Indeed, since every locally constant function on $[0,1]$ is in fact constant, this will imply $I(0) = I(1)$ as required.

Fix $s_0 \in [0,1]$ and consider a constructible primitive Φ of f along γ_{s_0}. Clearly, Φ comes with a partition $0 = t_0 < t_1 < \cdots < t_n = 1$, disks Δ^j containing $\gamma_{s_0}([t_j, t_{j+1}])$ and a holomorphic primitive F_j of f on Δ^j for $j = 0, \ldots, n-1$, with $F_j = F_{j+1}$ on $\Delta^j \cap \Delta^{j+1}$ for $j = 0, \ldots, n-2$.

Since Γ is continuous, it is uniformly continuous on $[0,1] \times [0,1]$, i.e., for every $\varepsilon > 0$ there exists a sufficiently small $\delta > 0$ such that one has $|\Gamma(t,s) - \Gamma(t',s')| < \varepsilon$ provided $|t - t'| < \delta$ and $|s - s'| < \delta$. In particular, for all $t \in [0,1]$ we have $|\Gamma(t,s) - \Gamma(t,s_0)| < \varepsilon$ if $|s - s_0| < \delta$. Choose ε small enough to guarantee that for the corresponding number δ one has $\Gamma(t,s) \subset \Delta^j$ if $t_j \leq t \leq t_{j+1}$, $|s - s_0| < \delta$, $j = 0, \ldots, n-1$ (see Fig. 9.1). One can now build a constructible primitive of f along γ_s, say Φ_s, by using the existing partition $0 = t_0 < t_1 < \cdots < t_n = 1$, the disks Δ^j and the holomorphic primitives F_j for $j = 0, \ldots, n-1$, which were used to produce the constructible primitive Φ along γ_{s_0}. Namely, if $t_j \leq t \leq t_{j+1}$, we set $\Phi_s(t) := F_j(\gamma_s(t))$ (clearly, $\Phi_{s_0} = \Phi$). We then have

$$I(s_0) = \Phi(1) - \Phi(0) = F_{n-1}(\gamma_{s_0}(1)) - F_0(\gamma_{s_0}(0)),$$

$$I(s) = \Phi_s(1) - \Phi_s(0) = F_{n-1}(\gamma_s(1)) - F_0(\gamma_s(0)). \tag{9.1}$$

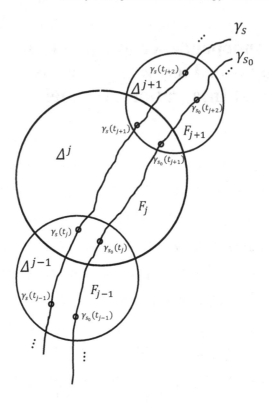

Fig. 9.1

Consider two cases.

Case 1. Suppose that γ and $\tilde{\gamma}$ are homotopic as paths with common endpoints. Then, since $\gamma_s(0) = \gamma_{s_0}(0)$ and $\gamma_s(1) = \gamma_{s_0}(1)$, it follows from (9.1) that $I(s) = I(s_0)$ for all $s \in [0,1]$ satisfying $|s - s_0| < \delta$. Hence $I(s)$ is locally constant on $[0,1]$ as required.

Case 2. Suppose that γ and $\tilde{\gamma}$ are homotopic as closed paths. Consider the difference $\psi := F_{n-1} - F_0$ on the intersection $\Delta^0 \cap \Delta^{n-1}$ (notice that $\Delta^0 \cap \Delta^{n-1}$ is non-empty and contains the points $z_1 := \gamma_{s_0}(0) = \gamma_{s_0}(1)$ and $z_2 := \gamma_s(0) = \gamma_s(1)$). Since $\psi \equiv C$ for some $C \in \mathbb{C}$ (cf. Step 3 in Lecture 7), we then see from (9.1) that

$$I(s_0) = F_{n-1}(z_1) - F_0(z_1) = \psi(z_1) = C, \quad I(s) = F_{n-1}(z_2) - F_0(z_2) = \psi(z_2) = C,$$

hence again we have $I(s) = I(s_0)$ for all $s \in [0,1]$ satisfying $|s - s_0| < \delta$.

The proof is complete. $\square$

Corollary 9.1. *Let $D \subset \mathbb{C}$ be a simply-connected domain and $f \in H(D)$. Then for any closed path γ in D one has $\int_\gamma f\,dz = 0$.*

Next, we will obtain the following generalisation of Theorem 7.1 to arbitrary simply-connected domains:

Corollary 9.2. *Let $D \subset \mathbb{C}$ be a simply-connected domain and $f \in H(D)$. Then f has a holomorphic primitive on D.*

Proof. Fix $a \in D$ and for $z \in D$ set

$$F(z) := \int_{\gamma(a,z)} f(\zeta)d\zeta,$$

where $\gamma(a,z)$ is *any* path in D joining a and z. By Theorem 9.1 (or, alternatively, by Corollary 9.1), the function F is well-defined, i.e., the above integral is independent of the choice of $\gamma(a,z)$. The proof that F' exists everywhere on D and is equal to f proceeds by using Corollary 9.1 and arguing as in the proof of Lemma 7.1 (provide full details!). $\square$

Next, we shall introduce an important class of paths.

Definition 9.1. A closed path γ in $\overline{\mathbb{C}}$ is called *a Jordan path* if it has no self-intersections other than at the endpoints, i.e., we have $\gamma(t_1) \neq \gamma(t_2)$ if $t_1 < t_2$ and $\{t_1, t_2\} \neq \{0, 1\}$.

We will now state without proof the following fact, which seems intuitively clear but is non-trivial to give a formal argument for:

Theorem 9.2. (The Jordan Curve Theorem) *Let γ be a Jordan path in $\mathbb{C}$ and $\Gamma := \gamma([0,1])$ its image. Then the complement $\mathbb{C} \setminus \Gamma$ is the union of two non-intersecting domains in $\mathbb{C}$, one bounded, the other unbounded, with Γ being the boundary of each domain.*

The Jordan paths that we consider below satisfy some conditions of differentiability and "smoothness".

Definition 9.2. A path γ in $\mathbb{C}$ is called *a piecewise C^1-path* if there exists a partition $0 = t_0 < t_1 < \cdots < t_n = 1$ of $[0,1]$ such that for every $j = 0, \ldots, n-1$ the restriction of γ to $[t_j, t_{j+1}]$ is a C^1-map, where the derivatives at t_j and t_{j+1} are understood as one-sided derivatives.

Remark 9.1. Let γ be a piecewise C^1-path in $\mathbb{C}$. The path γ comes with a partition $0 = t_0 < t_1 < \cdots < t_n = 1$ of $[0,1]$ as in Definition 9.2, and for every $j = 0, \ldots, n-1$ we consider the C^1-path $\gamma^j(t) := \gamma((1-t) \cdot t_j + t \cdot t_{j+1})$. We thus see that, loosely speaking, γ "consists" of finitely many C^1-paths "continuously joined together". Notice that for the path γ the partition $0 = t_0 < t_1 < \cdots < t_n = 1$ is *not* unique as one can split each of the segments $[t_j, t_{j+1}]$ into subsegments.

Definition 9.3. Let $D \subset \mathbb{C}$ be a domain, $f \in C(D)$ and γ a piecewise C^1-path in D. Define *the integral of f over γ* as

$$\int_\gamma f(z)dz := \sum_{j=0}^{n-1} \int_{\gamma^j} f(z)dz,$$

where the integrals in the right-hand side are understood in the sense of Definition 6.2.

The above definition is independent of the choice of a partition of $[0,1]$ (check!) and generalises Definition 6.2, in which only C^1-paths were allowed. Furthermore, if $f \in H(D)$, Definition 9.3 agrees with Definition 8.1 (check!).

Definition 9.4. *The length of a piecewise C^1-path γ is defined as*

$$|\gamma| := \sum_{j=0}^{n-1} \int_{t_j}^{t_{j+1}} |\gamma'(t)| dt = \sum_{j=0}^{n-1} |\gamma^j|.$$

This definition is independent of the choice of a partition of $[0,1]$, and Proposition 6.1 holds for piecewise C^1-paths (check!).

Remark 9.2. It is not hard to show that Properties 1–5 of integrals stated in Lecture 6 remain valid for the integral introduced in Definition 9.3, with ψ in Property 2 being, for instance, a monotone C^1-function and $|\gamma|$ in Property 5 understood in the sense of Definition 9.4 (check!).

For future reference, if $P, Q \in C(D)$ are real-valued functions, we also set

$$\int_\gamma P dx + Q dy := \sum_{j=0}^{n-1} \int_{\gamma^j} P dx + Q dy, \tag{9.2}$$

where the right-hand side contains integrals over C^1-paths of the kind studied in real analysis. Again, this definition is independent of the choice of a partition of $[0,1]$ (check!). In fact, in what follows we will allow P and Q to be complex-valued, with the integrals in the right-hand side of (9.2) made meaningful by separating the real and imaginary parts of the integrands. As in Lecture 6, for the integrals introduced in Definition 9.3 and formula (9.2) we have

$$\int_\gamma f dz = \int_\gamma u dx - v dy + i \int_\gamma v dx + u dy = \int_\gamma f dx + i f dy. \tag{9.3}$$

We shall now constrain Definition 9.2 a little.

Definition 9.5. A path γ in $\mathbb{C}$ is called *piecewise C^1-smooth* if there exists a partition $0 = t_0 < t_1 < \cdots < t_n = 1$ of $[0,1]$ such that for every $j = 0, \ldots, n-1$ the restriction of γ to $[t_j, t_{j+1}]$ is a C^1-map and its derivative is nowhere zero on $[t_j, t_{j+1}]$. If one can choose a partition with $n = 1$, the path is called C^1-*smooth*.

An example of the image of a piecewise C^1-smooth Jordan path γ is shown in Fig. 9.2 (here $n = 5$), where the spikes arise from the points at which the directions of the corresponding one-sided derivatives of γ differ. Notice, however, that some of the points in the image that appear to be "smooth" may also correspond to points in $[0,1]$ where γ' does not exist. For such points the directions of the one-sided derivatives of γ coincide whereas their values differ. For instance, let

$$f(t) := \begin{cases} \dfrac{t}{2} & \text{if } 0 \le t \le \dfrac{1}{2}, \\ \dfrac{3t}{2} - \dfrac{1}{2} & \text{if } \dfrac{1}{2} \le t \le 1 \end{cases}$$

and consider the piecewise C^1-smooth path $\gamma(t) := (1 - f(t)) + 2if(t)$. The one-sided derivatives of γ at the point $1/2$ are different but proportional, and the image of γ is just the segment with endpoints 1 and $2i$, which looks perfectly "smooth".

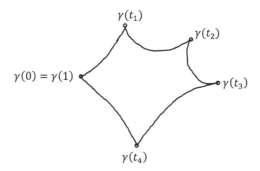

Fig. 9.2

Definition 9.6. A domain $D \subset \mathbb{C}$ is called *a Jordan domain* if its boundary ∂D in $\mathbb{C}$ can be represented as the disjoint union of the images of finitely many piecewise C^1-smooth Jordan paths.[1]

An example of a bounded Jordan domain is shown in Fig. 9.3 (see also the two domains making up the complement to the image of the path γ in Fig. 9.2).

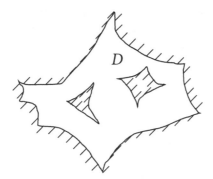

Fig. 9.3

In what follows, we will often deal with a function holomorphic on a neighbourhood of the boundary of a Jordan domain and integrate it over Jordan paths whose images yield the boundary components. Specifically, let Γ be a boundary component of a Jordan domain D and f be holomorphic on a neighbourhood of ∂D. Let, further, γ, $\tilde{\gamma}$ be any two piecewise C^1-smooth Jordan paths with image Γ (we say that each

---

[1] In the literature a more general definition is often used: a domain $D \subset \mathbb{C}$ is called Jordan if its boundary is homeomorphic to a circle. The definition given here suffices for our purposes.

of γ, $\tilde{\gamma}$ *parametrises* Γ). Consider the respective homeomorphisms $\hat{\gamma}, \widehat{\tilde{\gamma}}$, from the unit circle S^1 to Γ (see Remark 8.5) and write $\widehat{\tilde{\gamma}} = \hat{\gamma} \circ \Psi$, where $\Psi := \hat{\gamma}^{-1} \circ \widehat{\tilde{\gamma}}$ is a homeomorphism of S^1. Then by Remark 8.5 we observe that γ is homotopic either to $\tilde{\gamma}$ or to $\tilde{\gamma}_-$, where $\tilde{\gamma}_-(t) = \tilde{\gamma}(1-t)$. Notice that $\int_{\tilde{\gamma}_-} f \, dz = -\int_{\tilde{\gamma}} f \, dz$ (cf. Property 3 of integrals from Lecture 6). Therefore, by Theorem 9.1 we see that either $\int_{\gamma} f \, dz = \int_{\tilde{\gamma}} f \, dz$ or $\int_{\gamma} f \, dz = -\int_{\tilde{\gamma}} f \, dz$. Thus, for any path γ parametrising Γ the integral $\int_{\gamma} f \, dz$ takes one of only two possible values, and these values are opposite to each other. We will now make a canonical choice of one of the values and call it *the integral of f over* Γ, which we denote by $\int_{\Gamma} f(z) \, dz$.

Let $\gamma = (\gamma_1, \gamma_2) = \gamma_1 + i\gamma_2$ be any piecewise C^1-smooth Jordan path parametrising Γ. We need to decide which of the two values should be called the integral of f over Γ: $\int_{\gamma} f \, dz$ or $\int_{\gamma_-} f \, dz$. The path γ comes with a partition $0 = t_0 < t_1 < \cdots < t_n = 1$ of $[0,1]$. Fix $0 \leq j \leq n-1$ and consider $\tau_0 \in (t_j, t_{j+1})$. Then in a neighbourhood of the point $z_0 := \gamma(\tau_0)$ the component Γ can be written either as $y = g(x)$ or as $x = h(y)$ for some C^1-functions g, h. If Γ is given as $y = g(x)$, by Theorem 9.2 the domain D near z_0 is written as either $\Gamma_+ := \{z \in \mathbb{C} : y > g(x)\}$ or $\Gamma_- := \{z \in \mathbb{C} : y < g(x)\}$. If D is represented as Γ_+ and $\gamma_1'(\tau_0) > 0$ or D is represented as Γ_- and $\gamma_1'(\tau_0) < 0$, we set $\int_{\Gamma} f \, dz := \int_{\gamma} f \, dz$; otherwise define $\int_{\Gamma} f \, dz := \int_{\gamma_-} f \, dz = -\int_{\gamma} f \, dz$. If Γ is given as $x = h(y)$, then D near z_0 is written as either $\Gamma_+ := \{z \in \mathbb{C} : x > h(y)\}$ or $\Gamma_- := \{z \in \mathbb{C} : x < h(y)\}$. If D is represented as Γ_+ and $\gamma_2'(\tau_0) < 0$ or D is represented as Γ_- and $\gamma_2'(\tau_0) > 0$, we set $\int_{\Gamma} f \, dz := \int_{\gamma} f \, dz$; otherwise define $\int_{\Gamma} f \, dz := \int_{\gamma_-} f \, dz = -\int_{\gamma} f \, dz$. Notice that this definition is independent of the choice of a partition of $[0,1]$, as well as j and τ_0 (check!).

Intuitively, the above procedure for choosing between γ and γ_- to parametrise Γ means that we pick γ if the domain D "stays on the left as we move along γ" and pick γ_- otherwise. Fig. 9.4 shows these two possibilities: in part (a) the domain D "stays on the right", so γ needs to be reversed into γ_- whereas in part (b) the domain D "stays on the left" and reversing is not required.

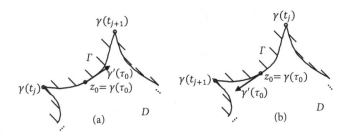

Fig. 9.4

The above choice of a parametrisation for Γ is sometimes said to *agree with the standard orientation on* Γ. Loosely speaking, we "orient" "the outer boundary component of ∂D" "anti-clockwise" and all "inner boundary components of ∂D" "clockwise" (see Fig. 9.5 where the arrows indicate the directions of the tangent vectors to paths parametrising the components of ∂D for the domain shown in Fig. 9.3).

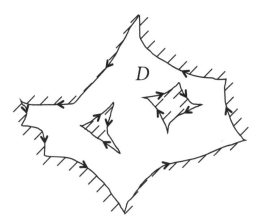

Fig. 9.5

Definition 9.7. Let D be a Jordan domain with $\partial D = \bigsqcup_{j=1}^{M} \Gamma^j$, where Γ^j is the image of a piecewise C^1-smooth Jordan path for every j, and let f be a function holomorphic on a neighbourhood of ∂D. Define the *integral of f over ∂D* as

$$\int_{\partial D} f(z)dz := \sum_{j=1}^{M} \int_{\Gamma^j} f(z)dz. \tag{9.4}$$

Note that Definition 9.7 agrees with the definition of the integral $\int_{\partial T} f dz$ over the boundary of a triangle in formula (7.1) (check!).

Suppose now that two complex-valued functions P and Q are continuous on a neighbourhood of the boundary of a Jordan domain D. Choose some parametrisations $\gamma^1, \ldots, \gamma^M$ of the boundary components $\Gamma^1, \ldots, \Gamma^M$ corresponding to the standard orientation and set

$$\int_{\partial D} Pdx + Qdy := \sum_{j=1}^{M} \int_{\gamma^j} Pdx + Qdy. \tag{9.5}$$

This definition is justified by the following fact:

Proposition 9.1. *Each of the integrals $\int_{\gamma^j} Pdx + Qdy$ in (9.5) is independent of the choice of γ^j.*

Proof. Homework. (Hint: use independence of re-parametrisation for the integral $\int_{\gamma^j} Pdx + Qdy$ similar to Property 2 in Lecture 6; take special care of the points where the paths involved are not continuously differentiable.) □

One can now set $\int_{\Gamma^j} P dx + Q dy := \int_{\gamma^j} P dx + Q dy$ for any choice of a parametrisation γ^j of Γ^j corresponding to the standard orientation on Γ^j. Then formula (9.5) takes a form analogous to (9.4):

$$\int_{\partial D} P dx + Q dy = \sum_{j=1}^{M} \int_{\Gamma^j} P dx + Q dy.$$

Exercises

9.1. Are the paths $\gamma(t) := e^{4\pi i t}$ and $\tilde{\gamma}(t) := e^{-2\pi i t}$ homotopic in $\mathbb{C} \setminus \{0\}$? Prove your conclusion.

9.2. Find the following integrals:

$$\text{(i)} \int_{|z|=1} \frac{\sin e^z}{z^3 - 9z^2 + 26z - 24} dz,$$

$$\text{(ii)} \int_{|z|=2} \frac{1}{z^{741} + 1} dz,$$

$$\text{(iii)} \int_{|z|=3} \frac{(z-1)e^{\sin z}}{z^3 - 10z^2 + 29z - 20} dz,$$

where $|z| = r$ is a shorthand for the path $re^{2\pi i t}$. (Hint: in Part (ii) consider "enlarging" the path of integration.)

9.3. Let $D \subset \mathbb{C}$ be a domain such that $\overline{\mathbb{C}} \setminus D$ is disconnected. Prove that there exist $f \in H(D)$ and a closed path γ in D such that $\int_\gamma f dz \neq 0$. (Hint: prove that one can find a non-empty subset C of the set of components of $\overline{\mathbb{C}} \setminus D$ such that: (i) every component in C lies in $\mathbb{C}$ and (ii) there exists a path γ in D that "surrounds" C).

9.4. Prove the necessity implication of Theorem 8.1.

9.5. Which of the following paths are piecewise C^1-paths:

$$\text{(i)} \ e^{2\pi i f(t)},$$
$$\text{(ii)} \ (1 - g(t)) + i g(t),$$
$$\text{(iii)} \ (2 - h(t)) + i h(t)?$$

Here f is the Cantor function,

$$g(t) := \begin{cases} t^2 & \text{if } 0 \leq t \leq \dfrac{1}{2}, \\ \dfrac{6t^3 + 1}{7} & \text{if } \dfrac{1}{2} \leq t \leq 1, \end{cases}$$

and

$$h(t) := \begin{cases} \sin \pi t & \text{if } 0 \le t \le \dfrac{1}{2}, \\[2mm] \dfrac{4t^2 + 2}{3} & \text{if } \dfrac{1}{2} \le t \le 1. \end{cases}$$

Prove your conclusions. Find the image of each path.

9.6. For every piecewise C^1-path γ from Exercise 9.5 find its length $|\gamma|$.

9.7. For every piecewise C^1-path γ from Exercise 9.5 find the integrals

$$\int_\gamma \bar{z}^2 dz \text{ and } \int_\gamma x dx + y^2 dy.$$

9.8. Which of the following paths are piecewise C^1-smooth:

$$\text{(i) } e^{16\pi i(t-1/2)^3},$$

$$\text{(ii) } e^{2\pi i t^2},$$

$$\text{(iii) } e^{\pi i f(t)},$$

$$\text{(iv) } (1 - g(t)) + ig(t),$$

$$\text{(v) } (1 - h(t)) + ih(t),$$

$$\text{(vi) } (2 - p(t)) + ip(t)?$$

Here

$$f(t) := \begin{cases} t & \text{if } 0 \le t \le \dfrac{1}{2}, \\[2mm] 2t - \dfrac{1}{2} & \text{if } \dfrac{1}{2} \le t \le 1, \end{cases}$$

$$g(t) := \begin{cases} t^3 & \text{if } 0 \le t \le \dfrac{1}{2}, \\[2mm] \dfrac{7t^2 - 1}{6} & \text{if } \dfrac{1}{2} \le t \le 1, \end{cases}$$

$$h(t) := \begin{cases} e^t & \text{if } 0 \le t \le \dfrac{1}{2}, \\[2mm] \dfrac{4(1 - \sqrt{e})t^2 - 1 + 4\sqrt{e}}{3} & \text{if } \dfrac{1}{2} \le t \le 1, \end{cases}$$

$$p(t) := \begin{cases} \sin \pi t & \text{if } 0 \le t \le \dfrac{1}{2}, \\[2mm] \dfrac{8t^3 + 6}{7} & \text{if } \dfrac{1}{2} \le t \le 1. \end{cases}$$

Prove your conclusions. Find the image of each path.

9.9. Suppose that a path γ is smooth at every $\tau_0 \in (0,1)$ in the sense of Definition 3.1. Does it follow that γ is C^1-smooth? If not, give examples of what exactly can go wrong.

9.10. Give an example of a closed C^1-smooth path γ for which the vectors $\gamma'(0)$ and $\gamma'(1)$ are not proportional.

9.11. Prove that the set

$$D := \{z \in \mathbb{C} : |z-1| + |z+1| < 4, \ |z| + |z-1| > 2, \ |z+1| > 1/4\}$$

is a Jordan domain and explicitly write a parametrisation of each boundary component Γ of D that agrees with the standard orientation on Γ.

9.12. Which of the following domains are Jordan:

(i) $\Delta \setminus [-1/2, 1/2]$,

(ii) $\Delta_3 \setminus \overline{\Delta}$,

(iii) $\mathbb{C} \setminus \mathbb{R}_+$,

(iv) $\mathbb{C} \setminus \{e^{\pi i t} : -1 \leq t \leq 1/2\}$,

(v) $\mathbb{C} \setminus (\overline{\Delta} \cup \overline{\Delta(2,1)})$,

(vi) $\mathbb{C} \setminus (\overline{\Delta} \cup \overline{\Delta(4,1)})$?

Prove your conclusions.

9.13. For every Jordan domain D from Exercise 9.12 explicitly write a parametrisation of each boundary component Γ that agrees with the standard orientation on Γ and compute the integrals

$$\int_{\partial D} e^{\frac{1}{z-2}} dz \quad \text{and} \quad \int_{\partial D} y dx + x^2 dy.$$

Lecture 10

Cauchy's Integral Theorem. Proof of Theorem 3.1. Cauchy's Integral Formula

We will now give further applications of Theorem 9.1. In what follows we often consider functions holomorphic on domains containing the closure $\overline{D}$ of a domain D. For such a function f we write $f \in H(\overline{D})$. Similarly, we write $f \in C^k(\overline{D})$ for a complex-valued function f of class C^k, with $k \geq 1$, on a domain containing $\overline{D}$.

The first important result based on Theorem 9.1 that will be obtained in this lecture is:

Theorem 10.1. (Cauchy's Integral Theorem) *Let $D \subset \mathbb{C}$ be a bounded Jordan domain and $f \in H(\overline{D})$. Then*

$$\int_{\partial D} f(z)dz = 0.$$

Proof. We will use the following fact from real analysis, which will be stated without proof:

Theorem 10.2. (Green's Theorem) *Let $D \subset \mathbb{C}$ be a bounded Jordan domain and $P, Q \in C^1(\overline{D})$. Then*

$$\int_{\partial D} Pdx + Qdy = \int_D \left(\frac{\partial Q}{\partial x} - \frac{\partial P}{\partial y} \right) dxdy,$$

with the integral in the left-hand side understood in the sense of formula (9.5) and the integral in the right-hand side made meaningful by separating the real and imaginary parts of the integrand.

Recall that by Theorem 3.1 (which is yet to be proved), one has $f \in C^1(\overline{D})$. Then formula (9.3) and Theorem 10.2 for $P := f$ and $Q := if$ imply

$$\int_{\partial D} f(z)dz = \int_D \left(i\frac{\partial f}{\partial x} - \frac{\partial f}{\partial y} \right) dxdy =$$
$$2i \int_D \frac{1}{2} \left(\frac{\partial f}{\partial x} + i\frac{\partial f}{\partial y} \right) dxdy = 2i \int_D \frac{\partial f}{\partial \bar{z}} dxdy = 0,$$

where the last equality is a consequence of the CR-equations (see Theorem 2.2). □

At first glance, it is not clear why Theorem 10.1 is claimed to have something to do with Theorem 9.1. Note, however, that the proof of Theorem 10.1 relies on Theorem 3.1, and it is the latter result that requires Theorem 9.1. It is now time for us to give a proof of this fact.

Proof (Theorem 3.1). First, we will obtain the following:

Theorem 10.3. (Cauchy's Integral Formula for a Disk) *If* $f \in H(\overline{\Delta(a,r)})$, *with* $0 < r < \infty$, *then for all* $z \in \Delta(a,r)$ *one has*

$$f(z) = \frac{1}{2\pi i} \int_{\partial\Delta(a,r)} \frac{f(\zeta)}{\zeta - z} d\zeta. \tag{10.1}$$

Proof. Without loss of generality we assume that $\Delta(a,r) = \Delta$. Let U be a domain containing $\overline{\Delta}$ on which f is holomorphic (in Fig. 10.1 the boundary of U is the most outer curve). Fix $z \in \Delta$; we may suppose that $z \in [0,1)$. For $\zeta \neq z$ define

$$g(\zeta) := \frac{f(\zeta)}{\zeta - z}. \tag{10.2}$$

Next, fix $\varepsilon > 0$, $\delta > 0$, with $\varepsilon < \delta < 1 - z$, and consider the bounded Jordan domain $D_{\delta,\varepsilon}$ shown in Fig. 10.1.

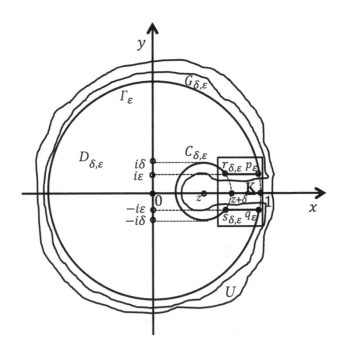

Fig. 10.1

The boundary of $D_{\delta,\varepsilon}$ consists of: an arc Γ_ε of S^1, an arc $C_{\delta,\varepsilon}$ of the circle of radius δ centred at z, a segment $[q_\varepsilon, s_{\delta,\varepsilon}]$ lying on the line $\{w \in \mathbb{C} : \text{Im } w = -\varepsilon\}$ and a segment $[r_{\delta,\varepsilon}, p_\varepsilon]$ lying on the line $\{w \in \mathbb{C} : \text{Im } w = \varepsilon\}$. We parametrise the unit circle S^1 as $e^{2\pi i t}$ (the anti-clockwise parametrisation) and the circle of radius δ centred at z as $z + \delta e^{-2\pi i t}$ (the clockwise parametrisation), where $0 \leq t \leq 1$. Together with the usual parametrisations of the segments $[q_\varepsilon, s_{\delta,\varepsilon}]$ and $[r_{\delta,\varepsilon}, p_\varepsilon]$, this yields a parametrisation, say $\gamma_{\delta,\varepsilon}$, of $\partial D_{\delta,\varepsilon}$, where we can assume that $\gamma_{\delta,\varepsilon}(0) = \gamma_{\delta,\varepsilon}(1) = p_\varepsilon$. Here for the paths $\gamma_{\delta,\varepsilon}^j$ (as defined in Remark 9.1) we have: $\gamma_{\delta,\varepsilon}^0$ corresponds to Γ_ε, $\gamma_{\delta,\varepsilon}^1$ to $[q_\varepsilon, s_{\delta,\varepsilon}]$, $\gamma_{\delta,\varepsilon}^2$ to $C_{\delta,\varepsilon}$, $\gamma_{\delta,\varepsilon}^3$ to $[r_{\delta,\varepsilon}, p_\varepsilon]$, and

$$\int_{\partial D_{\delta,\varepsilon}} g d\zeta = \int_{\gamma_{\delta,\varepsilon}} g d\zeta = \int_{\gamma_{\delta,\varepsilon}^0} g d\zeta + \int_{\gamma_{\delta,\varepsilon}^1} g d\zeta + \int_{\gamma_{\delta,\varepsilon}^2} g d\zeta + \int_{\gamma_{\delta,\varepsilon}^3} g d\zeta.$$

We now slightly enlarge $D_{\delta,\varepsilon}$ to obtain a simply-connected domain $G_{\delta,\varepsilon}$ containing $\overline{D}_{\delta,\varepsilon}$, lying in U, and such that $g \in H(G_{\delta,\varepsilon})$. In Fig. 10.1 the (connected) boundary of $G_{\delta,\varepsilon}$ sits between that of $D_{\delta,\varepsilon}$ and that of U. Then by Corollary 9.1 we see $\int_{\gamma_{\delta,\varepsilon}} g d\zeta = 0$.

Let $\varepsilon \to 0$. One can show

$$\int_{\gamma_{\delta,\varepsilon}^1} g d\zeta + \int_{\gamma_{\delta,\varepsilon}^3} g d\zeta \to 0,$$

$$\int_{\gamma_{\delta,\varepsilon}^0} g d\zeta \to \int_{|\zeta|=1} g d\zeta, \tag{10.3}$$

$$\int_{\gamma_{\delta,\varepsilon}^2} g d\zeta \to -\int_{|\zeta-z|=\delta} g d\zeta,$$

where both the unit circle S^1 and the circle of radius δ centred at z are now parametrised in the anti-clockwise way (here and below, when we write $|\zeta - a| = r$ as a path of integration, we mean the path $a + re^{2\pi i t}$). We will check the first claim in (10.3) and leave the remaining two to the reader since the proofs are similar.

Write

$$\int_{\gamma_{\delta,\varepsilon}^1} g d\zeta + \int_{\gamma_{\delta,\varepsilon}^3} g d\zeta =$$
$$\int_0^1 g((1-t)q_\varepsilon + ts_{\delta,\varepsilon})(s_{\delta,\varepsilon} - q_\varepsilon) dt + \int_0^1 g((1-t)r_{\delta,\varepsilon} + tp_\varepsilon)(p_\varepsilon - r_{\delta,\varepsilon}) dt.$$

When $\varepsilon \to 0$, we have $p_\varepsilon, q_\varepsilon \to 1$, $r_{\delta,\varepsilon}, s_{\delta,\varepsilon} \to z + \delta$. Clearly, there exists a compact subset $\mathbf{K} \subset U$ containing the segment $[z + \delta, 1]$ as well as the segments $[q_\varepsilon, s_{\delta,\varepsilon}]$ and $[r_{\delta,\varepsilon}, p_\varepsilon]$ for all sufficiently small ε, on which the function g is defined and continuous, hence uniformly continuous (in Fig. 10.1 the set $\mathbf{K}$ is a rectangle). By the uniform continuity of g on $\mathbf{K}$ we then see that, as $\varepsilon \to 0$, the families $g((1-t)q_\varepsilon + ts_{\delta,\varepsilon})(s_{\delta,\varepsilon} - q_\varepsilon)$ and $g((1-t)r_{\delta,\varepsilon} + tp_\varepsilon)(p_\varepsilon - r_{\delta,\varepsilon})$ converge to $g((1-t) + t(z+\delta))(z + \delta - 1)$ and $g((1-t)(z+\delta) + t)(1 - z - \delta)$, respectively, *uniformly on* $[0, 1]$. Therefore, as $\varepsilon \to 0$, we have (cf. Lemma 11.1 in Lecture 11)

$$\int_0^1 g((1-t)q_\varepsilon + ts_{\delta,\varepsilon})(s_{\delta,\varepsilon} - q_\varepsilon)dt \to$$

$$\int_0^1 g((1-t)+t(z+\delta))(z+\delta-1)dt = \int_{[1,z+\delta]} gd\zeta,$$

and

$$\int_0^1 g((1-t)r_{\delta,\varepsilon} + tp_\varepsilon)(p_\varepsilon - r_{\delta,\varepsilon}) \to$$

$$\int_0^1 g((1-t)(z+\delta)+t)(1-z-\delta)dt = \int_{[z+\delta,1]} gd\zeta = -\int_{[1,z+\delta]} gd\zeta,$$

which establishes the first claim in (10.3).

By (10.3) one has

$$\int_{|\zeta-z|=\delta} gd\zeta = \int_{|\zeta|=1} gd\zeta. \tag{10.4}$$

Theorem 10.3 now follows from formula (10.4) and the lemma stated below by letting $\delta \to 0$. $\square$

Lemma 10.1. *When* $\delta \to 0$, *we have*

$$\int_{|\zeta-z|=\delta} gd\zeta \to 2\pi i f(z).$$

Proof. Represent the number $2\pi i$ as

$$2\pi i = \int_{|\zeta-z|=\delta} \frac{1}{\zeta-z} d\zeta$$

(see Example 6.1 for $m = 1$). Then

$$\int_{|\zeta-z|=\delta} gd\zeta - 2\pi i f(z) = \int_{|\zeta-z|=\delta} gd\zeta - \int_{|\zeta-z|=\delta} \frac{f(z)}{\zeta-z} d\zeta = \int_{|\zeta-z|=\delta} \frac{f(\zeta)-f(z)}{\zeta-z} d\zeta.$$

Therefore

$$\left| \int_{|\zeta-z|=\delta} gd\zeta - 2\pi i f(z) \right| \le \frac{\max_{|\zeta-z|=\delta} |f(\zeta)-f(z)|}{\delta} 2\pi\delta = 2\pi \max_{|\zeta-z|=\delta} |f(\zeta)-f(z)|,$$

and the continuity of f at the point z yields the result. $\square$

To finalise the proof of Theorem 3.1, for every $z_0 \in D$ find a disk $\Delta(z_0, \delta)$, with closure lying in D, and apply formula (10.1) to $\Delta(z_0, \delta)$ and the function at hand. The lemma stated below now implies that the function is infinitely many times $\mathbb{C}$-differentiable on $\Delta(z_0, \delta)$. As z_0 is an arbitrary point of D, this completes the proof of Theorem 3.1. $\square$

---

[1] For pedagogical reasons, we gave a longer proof of this identity here (cf. Exercise 10.2). A shorter one can be deduced directly from Theorem 9.1 (see Exercise 10.3).

Lemma 10.2. *Let f be a function in $C(\overline{\Delta(a,r)})$, with $0 < r < \infty$, for which formula (10.1) holds. Then f is infinitely many times $\mathbb{C}$-differentiable on $\Delta(a,r)$, with the kth derivative $f^{(k)}$ found from the formula*

$$f^{(k)}(z) = \frac{k!}{2\pi i} \int_{\partial\Delta(a,r)} \frac{f(\zeta)}{(\zeta-z)^{k+1}} d\zeta \quad \forall z \in \Delta(a,r), \forall k \in \mathbb{N}. \tag{10.5}$$

Proof. We will prove the following statement by induction on $k \geq 1$: the function f is k times $\mathbb{C}$-differentiable on $\Delta(a,r)$ and $f^{(k)}$ is given by formula (10.5). For $k = 1$ we need to show that $f \in H(\Delta(a,r))$ and

$$f'(z) = \frac{1}{2\pi i} \int_{\partial\Delta(a,r)} \frac{f(\zeta)}{(\zeta-z)^2} d\zeta \quad \forall z \in \Delta(a,r). \tag{10.6}$$

Fix $z \in \Delta(a,r)$ and using formula (10.1) for all sufficiently small Δz write

$$\left| \frac{f(z+\Delta z) - f(z)}{\Delta z} - \frac{1}{2\pi i} \int_{\partial\Delta(a,r)} \frac{f(\zeta)}{(\zeta-z)^2} d\zeta \right| =$$

$$\left| \frac{1}{2\pi i\Delta z} \int_{\partial\Delta(a,r)} f(\zeta) \left(\frac{1}{\zeta-z-\Delta z} - \frac{1}{\zeta-z} \right) d\zeta - \frac{1}{2\pi i} \int_{\partial\Delta(a,r)} \frac{f(\zeta)}{(\zeta-z)^2} d\zeta \right| =$$

$$\left| \frac{1}{2\pi i} \int_{\partial\Delta(a,r)} f(\zeta) \left(\frac{1}{(\zeta-z-\Delta z)(\zeta-z)} - \frac{1}{(\zeta-z)^2} \right) d\zeta \right| \leq$$

$$\frac{1}{2\pi} \max_{|\zeta-a|=r} |f(\zeta)| \max_{|\zeta-a|=r} \left| \frac{1}{(\zeta-z-\Delta z)(\zeta-z)} - \frac{1}{(\zeta-z)^2} \right| 2\pi r.$$

Since $|f(\zeta)|$ is bounded on $\partial\Delta(a,r)$, it follows that the expression in the last line of the above formula tends to zero as $\Delta z \to 0$. This shows that $f \in H(\Delta(a,r))$ and proves formula (10.6).

Let now $k > 1$ and assume that f is $k - 1$ times $\mathbb{C}$-differentiable on $\Delta(a,r)$ with

$$f^{(j)}(z) = \frac{j!}{2\pi i} \int_{\partial\Delta(a,r)} \frac{f(\zeta)}{(\zeta-z)^{j+1}} d\zeta \quad \forall z \in \Delta(a,r), j = 1, \ldots, k-1. \tag{10.7}$$

We will prove that f is k times $\mathbb{C}$-differentiable on $\Delta(a,r)$ by differentiating the right-hand side of (10.7) for $j = k - 1$. This will yield that the derivative $(f^{(k-1)})'$ exists and is equal to the expression in the right-hand side of (10.5).

Fix $z \in \Delta(a,r)$ and for all sufficiently small Δz write

$$\left| \frac{f^{(k-1)}(z+\Delta z) - f^{(k-1)}(z)}{\Delta z} - \frac{k!}{2\pi i} \int_{\partial\Delta(a,r)} \frac{f(\zeta)}{(\zeta-z)^{k+1}} d\zeta \right| =$$

$$\left| \frac{(k-1)!}{2\pi i\Delta z} \int_{\partial\Delta(a,r)} f(\zeta) \left(\frac{1}{(\zeta-z-\Delta z)^k} - \frac{1}{(\zeta-z)^k} \right) d\zeta - \frac{k!}{2\pi i} \int_{\partial\Delta(a,r)} \frac{f(\zeta)}{(\zeta-z)^{k+1}} d\zeta \right| =$$

$$\left| \frac{(k-1)!}{2\pi i \Delta z} \int_{\partial \Delta(a,r)} f(\zeta) \frac{(\zeta - z)^k - (\zeta - z - \Delta z)^k}{(\zeta - z - \Delta z)^k (\zeta - z)^k} d\zeta - \frac{k!}{2\pi i} \int_{\partial \Delta(a,r)} \frac{f(\zeta)}{(\zeta - z)^{k+1}} d\zeta \right| =$$

$$\left| \frac{(k-1)!}{2\pi i} \int_{\partial \Delta(a,r)} f(\zeta) \frac{\sum_{j=0}^{k-1} (-1)^{k-j-1} \binom{k}{j} (\zeta - z)^j \Delta z^{k-j-1}}{(\zeta - z - \Delta z)^k (\zeta - z)^k} d\zeta - \right.$$
$$\left. \frac{k!}{2\pi i} \int_{\partial \Delta(a,r)} \frac{f(\zeta)}{(\zeta - z)^{k+1}} d\zeta \right| \leq$$

$$\left| \frac{(k-1)!}{2\pi i} \int_{\partial \Delta(a,r)} f(\zeta) \frac{\sum_{j=0}^{k-2} (-1)^{k-j-1} \binom{k}{j} (\zeta - z)^j \Delta z^{k-j-1}}{(\zeta - z - \Delta z)^k (\zeta - z)^k} d\zeta \right| +$$

$$\left| \frac{k!}{2\pi i} \int_{\partial \Delta(a,r)} f(\zeta) \left(\frac{1}{(\zeta - z - \Delta z)^k (\zeta - z)} - \frac{1}{(\zeta - z)^{k+1}} \right) d\zeta \right| \leq$$

$$\frac{(k-1)!}{2\pi} \sum_{j=0}^{k-2} \binom{k}{j} |\Delta z|^{k-j-1} \left| \int_{\partial \Delta(a,r)} \frac{f(\zeta)}{(\zeta - z - \Delta z)^k (\zeta - z)^{k-j}} d\zeta \right| +$$

$$\frac{k!}{2\pi} \max_{|\zeta - a| = r} |f(\zeta)| \max_{|\zeta - a| = r} \left| \frac{1}{(\zeta - z - \Delta z)^k (\zeta - z)} - \frac{1}{(\zeta - z)^{k+1}} \right| 2\pi r \leq$$

$$(k-1)! r \sum_{j=0}^{k-2} \binom{k}{j} |\Delta z|^{k-j-1} \max_{|\zeta - a| = r} |f(\zeta)| \max_{|\zeta - a| = r} \frac{1}{|\zeta - z - \Delta z|^k |\zeta - z|^{k-j}} +$$

$$k! r \max_{|\zeta - a| = r} |f(\zeta)| \max_{|\zeta - a| = r} \left| \frac{1}{(\zeta - z - \Delta z)^k (\zeta - z)} - \frac{1}{(\zeta - z)^{k+1}} \right|.$$

It is clear that each of the two summands in the last expression tends to zero as $\Delta z \to 0$. This proves the lemma. $\quad \square$

We have thus fully established Theorem 3.1 and therefore Theorem 10.1. Before proceeding further, we record a consequence of Theorem 10.3 and Lemma 10.2.

Corollary 10.1. *Let f be as in Theorem* 10.3. *Then for $k = 0, 1, \ldots$ one has*

$$|f^{(k)}(a)| \leq \frac{k! \max_{|z-a|=r} |f(z)|}{r^k}.$$

Proof. Homework. $\quad \square$

We will now obtain a generalisation of Theorem 10.3 to arbitrary bounded Jordan domains.

Theorem 10.4. (Cauchy's Integral Formula) *Let $D \subset \mathbb{C}$ be a bounded Jordan domain and $f \in H(\overline{D})$. Then*

$$f(z) = \frac{1}{2\pi i} \int_{\partial D} \frac{f(\zeta)}{\zeta - z} d\zeta \ \forall z \in D. \tag{10.8}$$

Furthermore,

$$f^{(k)}(z) = \frac{k!}{2\pi i} \int_{\partial D} \frac{f(\zeta)}{(\zeta - z)^{k+1}} d\zeta \ \forall z \in D, \ \forall k \in \mathbb{N}.$$

Remark 10.1. The above theorem is truly remarkable as it states, in particular, that a function holomorphic on a bounded Jordan domain can be fully reconstructed from its values on the boundary.

Proof (Theorem 10.4). Fix $z \in D$ and a disk $\Delta(z, \delta)$ with closure lying in D. Consider the bounded Jordan domain $D_\delta := D \setminus \overline{\Delta(z, \delta)}$. For the domain in Fig. 9.5 the domain D_δ is shown in Fig. 10.2, where, as before, the arrows indicate the directions of the tangent vectors to paths parametrising the components of ∂D_δ.

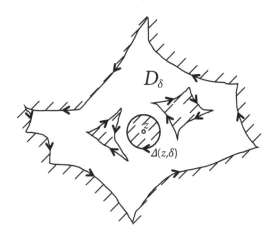

Fig. 10.2

Consider the function g defined by formula (10.2). Clearly, $g \in H(\overline{D_\delta})$. By Theorem 10.1 we have $\int_{\partial D_\delta} g(\zeta) d\zeta = 0$, which is equivalent to

$$\int_{|\zeta - z| = \delta} g d\zeta = \int_{\partial D} g d\zeta. \tag{10.9}$$

The first statement of Theorem 10.4 now follows from Lemma 10.1 and formula (10.9) by letting $\delta \to 0$. The second statement is derived from formula (10.8) by arguing as in the proof of Lemma 10.2. $\quad\square$

Exercises

10.1. Assuming that D is a polygon, deduce Theorem 10.1 from Lemma 7.2.

10.2. Explain how one can attempt to prove Theorem 10.1 by appealing to Corollary 9.1 without referring to Theorem 10.2. (Hint: consider making "cuts" in the domain in the spirit of the proof of Theorem 10.3.)

10.3. Deduce identity (10.4) directly from Theorem 9.1. (Hint: notice that the two circles are homotopic in a domain where the function g is holomorphic.)

10.4. Does the proof of Lemma 10.1 rely on the fact that f is holomorphic or a weaker condition on f suffices? If so, what is that condition? Prove your conclusion.

10.5. Let $D \subset \mathbb{C}$ be a domain, $f \in H(D)$ and assume that there exists $M > 0$ with $|f(z)| \le M$ for all $z \in D$. Prove that

$$|f'(z)| \le \frac{M}{\text{dist}(z, \partial D)} \quad \forall z \in D.$$

10.6. Let $D := \{z \in \mathbb{C} : \text{Re } z > 0\}$, $f \in H(D)$, and assume that there exits $M > 0$ such that $|f(z)| \le M$ for all $z \in D$. Prove that $\lim_{x \to +\infty} f'(x) = 0$, where the limit is taken over the positive direction of the x-axis.

10.7. Let

$$D := \left\{ z \in \mathbb{C} : 1 < x < \frac{11}{3}, -4 < y < 4 \right\} \setminus \left(\overline{\Delta(2, 1/2)} \cup \overline{\Delta(3, 1/3)} \right).$$

Prove that D is a Jordan domain and compute the integrals

(i) $\displaystyle \int_{\partial D} \frac{e^z \sin(z^2 + 2z - 1)}{z^3 - 9z^2 + 26z - 24} dz,$

(ii) $\displaystyle \int_{\partial D} \frac{z}{(z^2 - 5z + 6)(z - 2 - 3i)} dz,$

(iii) $\displaystyle \int_{\partial D} \frac{z^2}{(z^2 - 7z + 12)(z - 3 + 3i)^2} dz,$

(iv) $\displaystyle \int_{\partial D} e^x(x^3 + 2x^2 + 4)dx + \sin y(y^5 + 4y^3 + 1)dy,$

(v) $\displaystyle \int_{\partial D} x\cos(x^2 + y^2)e^{\sin(x^2+y^2)}dx + y\cos(x^2 + y^2)e^{\sin(x^2+y^2)}dy.$

10.8. Compute the right-hand side of formula (10.8) for $z \in \mathbb{C} \setminus \overline{D}$.

10.9. Let $f \in H(\overline{\Delta_r})$ for some $0 < r < \infty$. Prove that

$$f(0) = \frac{1}{2\pi} \int_0^{2\pi} f(re^{it})dt.$$

10.10. Let $D \subset \mathbb{C}$ be an unbounded Jordan domain and $f \in H(\overline{D})$. Suppose that f has a limit as $|z| \to \infty$, i.e., one can find $A \in \mathbb{C}$, called *the limit of f at* ∞, such that for every $\varepsilon > 0$ there exists a sufficiently large $R > 0$ with $|f(z) - A| < \varepsilon$ provided $|z| > R$. Prove that for every $z \in D$ the following holds:

$$f(z) = \frac{1}{2\pi i} \int_{\partial D} \frac{f(\zeta)}{\zeta - z} d\zeta + A.$$

10.11. For the function f introduced in Exercise 2.23 by formula (2.3) compute $f^{(k)}(z)$ for all $z \in \mathbb{C} \setminus [0, 1]$ and $k \in \mathbb{N}$.

Lecture 11

Morera's Theorem. Sequences and Series of Functions. Uniform Convergence Inside a Domain. Power Series. Abel's Theorem. Disk of Convergence. Radius of Convergence

Now that we have proved Theorem 3.1, we can establish the converse to Lemma 7.2, which will be useful for us in what follows.

Theorem 11.1. (Morera's Theorem) *Let $D \subset \mathbb{C}$ be a domain and $f \in C(D)$. Suppose that for any triangle T lying in D we have $\int_{\partial T} f \, dz = 0$. Then $f \in H(D)$.*

Proof. By Lemma 7.1, the function f has a holomorphic primitive on any disk contained in D. Using Theorem 3.1 and differentiating such locally defined primitives twice, we then see that f' exists everywhere on D as required (cf. Exercise 3.7). □

Next, we will discuss a connection between holomorphic functions and power series. Let us start by recalling a few fundamental facts on convergence of sequences and series of functions. All the functions are assumed to be $\mathbb{C}$-valued and defined on subsets of the complex plane, but the proofs of most statements given below are analogous to those of their counterparts in the real case, so we omit them. Note that many of the statements are special cases of familiar facts from metric space theory. Usually, the terms in a sequence or a series will be enumerated by the elements of the set $\{0\} \cup \mathbb{N} = \{0, 1, 2, \ldots\}$.

Before considering sequences and series of arbitrary functions, we will briefly discuss sequences and series of complex numbers.

Definition 11.1. A sequence $\{z_n\}$ of complex numbers *is convergent*, or *converges*, if there exists a complex number w, called *the limit of* $\{z_n\}$, such that for every $\varepsilon > 0$ there is $N \in \mathbb{N}$ with $|z_n - w| < \varepsilon$ for all $n \geq N$. In this case we write $w = \lim_{n \to \infty} z_n$ and say that the sequence $\{z_n\}$ *converges to* w. If $\{z_n\}$ does not converge, we say that the sequence *is divergent*, or *diverges*.

As the following important theorem shows, in order to prove that a sequence is convergent, one does not in fact need to guess what its limit is:

Theorem 11.2. (Cauchy's Criterion) *A sequence $\{z_n\}$ of complex numbers converges if and only if for every $\varepsilon > 0$ there exists $N \in \mathbb{N}$ such that $|z_n - z_m| < \varepsilon$ for all $n, m \geq N$.*

We shall now turn to series of complex numbers.

Definition 11.2. A series $\sum_{n=0}^{\infty} z_n$ of complex numbers *is convergent*, or *converges*, if *the sequence of its partial sums* $\{s_k := \sum_{n=0}^{k} z_n\}$ converges. In this case, the limit of the sequence, say s, is called *the sum of* $\sum_{n=0}^{\infty} z_n$; we write $s = \sum_{n=0}^{\infty} z_n$ and say that the series $\sum_{n=0}^{\infty} z_n$ *converges to* s. Further, we say that the series $\sum_{n=0}^{\infty} z_n$ *converges absolutely* if the series $\sum_{n=0}^{\infty} |z_n|$ converges. If $\sum_{n=0}^{\infty} z_n$ does not converge, we say that the series *is divergent*, or *diverges*.

It is clear that for series Theorem 11.2 becomes:

Theorem 11.3. (Cauchy's Criterion) *A series* $\sum_{n=0}^{\infty} z_n$ *of complex numbers converges if and only if for every* $\varepsilon > 0$ *there exists* $N \in \mathbb{N}$ *such that* $|\sum_{n=\ell}^{m} z_n| < \varepsilon$ *for all* $\ell, m \geq N$ *with* $m \geq \ell$.

Theorem 11.3 immediately implies:

Corollary 11.1. *If a series of complex numbers converges absolutely, then it converges.*

When a series of complex numbers $\sum_{n=0}^{\infty} z_n$ converges absolutely (hence converges), we often say that *the convergence [of* $\sum_{n=0}^{\infty} z_n$*] is absolute.*

Below we will need the following important fact:

Theorem 11.4.

(1) *If a series of complex numbers* $\sum_{n=0}^{\infty} z_n$ *converges absolutely, then any rearrangement of* $\sum_{n=0}^{\infty} z_n$ *converges to the same value;*

(2) *if a series of complex numbers* $\sum_{n=0}^{\infty} z_n$ *converges absolutely, then for any splitting of* $\sum_{n=0}^{\infty} z_n$ *into countably many mutually exclusive subseries, each of the subseries converges absolutely and the sum of the sums of the subseries, calculated for any ordering the subseries, is equal to the sum of* $\sum_{n=0}^{\infty} z_n$*;*

(3) *if two series of complex numbers* $\sum_{n=0}^{\infty} z_n$ *and* $\sum_{k=0}^{\infty} w_k$ *converge absolutely, then the series that consists of all pairwise products* $z_n w_k$ *converges to the product of the sums of* $\sum_{n=0}^{\infty} z_n$ *and* $\sum_{k=0}^{\infty} w_k$ *for any ordering of its terms;*

(4) *if* $\{a_{nk}\}$, $n, k = 0, 1, 2, \ldots$, *is a collection of complex numbers such that the limit*

$$\lim_{N \to \infty} \sum_{n=0}^{N} \left(\lim_{K \to \infty} \sum_{k=0}^{K} |a_{nk}| \right)$$

exists and is finite, then the limits

$$\lim_{N \to \infty} \sum_{n=0}^{N} \left(\lim_{K \to \infty} \sum_{k=0}^{K} a_{nk} \right) \quad \text{and} \quad \lim_{K \to \infty} \sum_{k=0}^{K} \left(\lim_{N \to \infty} \sum_{n=0}^{N} a_{nk} \right)$$

exist, are equal, and coincide with the sum of the double series $\sum_{n,k=0}^{\infty} a_{nk}$ *calculated for any ordering of its terms.*

Further, a useful consequence of Theorem 11.3 is:

Corollary 11.2. (Weierstrass' M-Test) *If for a series $\sum_{n=0}^{\infty} z_n$ of complex numbers there exists a convergent series of non-negative real numbers $\sum_{n=0}^{\infty} a_n$ such that $|z_n| \leq a_n$ for all n, then $\sum_{n=0}^{\infty} z_n$ converges absolutely.*

We now turn to sequences and series of complex-valued functions.

Definition 11.3. A sequence $\{f_n\}$ of complex-valued functions defined on a subset $S \subset \mathbb{C}$ *is convergent at $z_0 \in S$, or converges at z_0,* if the sequence $\{f_n(z_0)\}$ converges. If $w := \lim_{n \to \infty} f_n(z_0)$, we say that the sequence $\{f_n\}$ *converges to w at z_0* and that *w is the limit of $\{f_n\}$ at z_0.* The sequence $\{f_n\}$ is said to be *divergent at z_0,* or to *diverge at z_0,* if the sequence $\{f_n(z_0)\}$ diverges.

Definition 11.4. A sequence $\{f_n\}$ of complex-valued functions defined on a subset $S \subset \mathbb{C}$ *is convergent on S, or converges on S,* if it converges at every point of S. In this case, the function $f(z) := \lim_{n \to \infty} f_n(z)$, with $z \in S$, is called *the limit of $\{f_n\}$;* we write $f = \lim_{n \to \infty} f_n$ and say that the sequence $\{f_n\}$ *converges to f [on S].* Further, we say that *the convergence [of $\{f_n\}$ to f] is uniform on a subset $S' \subset S$,* or that *the sequence converges [to f] uniformly on S',* or that *f is the uniform limit of the sequence on S',* if for every $\varepsilon > 0$ there exists $N \in \mathbb{N}$ such that $|f_n(z) - f(z)| < \varepsilon$ for all $n \geq N$ and all $z \in S'$.

The analogue of Theorem 11.2 for sequences of functions is:

Theorem 11.5. (Cauchy's Criterion) *A sequence $\{f_n\}$ of complex-valued functions defined on a subset $S \subset \mathbb{C}$ converges on S if and only if for every $\varepsilon > 0$ and $z \in S$ there exists $N \in \mathbb{N}$ such that $|f_n(z) - f_m(z)| < \varepsilon$ for all $n, m \geq N$. The sequence $\{f_n\}$ converges uniformly on a subset $S' \subset S$ if and only if for every $\varepsilon > 0$ there exists $N \in \mathbb{N}$ such that $|f_n(z) - f_m(z)| < \varepsilon$ for all $n, m \geq N$ and all $z \in S'$.*

Next, we consider series of functions.

Definition 11.5. A series $\sum_{n=0}^{\infty} f_n$ of complex-valued functions defined on a subset $S \subset \mathbb{C}$ *is convergent at $z_0 \in S$, or converges at z_0,* if the sequence of its partial sums $\{F_k := \sum_{n=0}^{k} f_n\}$ converges at z_0. If $s := \lim_{k \to \infty} F_k(z_0) = \sum_{n=0}^{\infty} f_n(z_0)$, we say that the series $\sum_{n=0}^{\infty} f_n$ *converges to s at z_0* and that *s is the sum of $\sum_{n=0}^{\infty} f_n$ at z_0.* The series is said to be *divergent at z_0,* or to *diverge at z_0,* if the sequence $\{F_k\}$ diverges at z_0. The series *is convergent on S, or converges on S,* if $\{F_k\}$ converges on S. In this case, the limit of the sequence, say a function F, is called *the sum of $\sum_{n=0}^{\infty} f_n$;* we write $F = \sum_{n=0}^{\infty} f_n$ and say that the series $\sum_{n=0}^{\infty} f_n$ *converges to F [on S].* Further, we say that *the convergence [of $\sum_{n=0}^{\infty} f_n$ to F] is uniform on a subset $S' \subset S$,* or that *the series converges [to F] uniformly on S',* or that *F is the uniform sum of the series on S',* if the sequence $\{F_k\}$ converges to F uniformly on S'. Also, we say that the series $\sum_{n=0}^{\infty} f_n$ *converges absolutely at $z_0 \in S$* if the series $\sum_{n=0}^{\infty} |f_n(z_0)|$ converges and that it *converges absolutely on S* if the series $\sum_{n=0}^{\infty} |f_n|$ converges on S.

For series of functions Theorem 11.5 turns into:

Theorem 11.6. (Cauchy's Criterion) *A series $\sum_{n=0}^{\infty} f_n$ of complex-valued functions defined on a subset $S \subset \mathbb{C}$ converges on S if and only if for every $\varepsilon > 0$ and $z \in S$ there exists $N \in \mathbb{N}$ such that $|\sum_{n=\ell}^{m} f_n(z)| < \varepsilon$ for all $\ell, m \geq N$ with $m \geq \ell$. The series $\sum_{n=0}^{\infty} f_n$ converges uniformly on a subset $S' \subset S$ if and only if for every $\varepsilon > 0$ there exists $N \in \mathbb{N}$ such that $|\sum_{n=\ell}^{m} f_n(z)| < \varepsilon$ for all $\ell, m \geq N$, with $m \geq \ell$, and all $z \in S'$.*

Analogously to Corollary 11.1 we have:

Corollary 11.3. *If a series of complex-valued functions defined on a subset $S \subset \mathbb{C}$ converges absolutely at $z_0 \in S$ (resp. on S), then it converges at z_0 (resp. on S).*

When a series $\sum_{n=0}^{\infty} f_n$ of complex-valued functions defined on a subset $S \subset \mathbb{C}$ converges absolutely at a point $z_0 \in S$ (resp. on S), we often say that *the convergence* [*of* $\sum_{n=0}^{\infty} f_n$] *is absolute at z_0* (resp. *on S*).

The analogue of Corollary 11.2 is:

Corollary 11.4. (Weierstrass' M-Test) *If for a series $\sum_{n=0}^{\infty} f_n$ of complex-valued functions defined on a subset $S \subset \mathbb{C}$ there exists a convergent series of non-negative real numbers $\sum_{n=0}^{\infty} a_n$ such that $|f_n(z)| \leq a_n$ for all n and $z \in S' \subset S$, then $\sum_{n=0}^{\infty} f_n$ converges absolutely and uniformly on the subset S'.*

For sequences and series of functions defined on domains we will utilise the following notion of convergence:

Definition 11.6. A sequence $\{f_n\}$ of complex-valued functions defined on a domain $D \subset \mathbb{C}$ is said to *converge uniformly inside D* if $\{f_n\}$ converges uniformly on every compact subset of D. In this case, if $f := \lim_{n \to \infty} f_n$, we say that *the convergence* [*of* $\{f_n\}$ *to* f] *is uniform inside D*, or that $\{f_n\}$ *converges to f uniformly inside D*. A series $\sum_{n=0}^{\infty} f_n$ of complex-valued functions defined on a domain $D \subset \mathbb{C}$ is said to *converge uniformly inside D* if the sequence of partial sums of the series converges uniformly inside D. In this case, if $F := \sum_{n=0}^{\infty} f_n$, we say that *the convergence* [*of* $\sum_{n=0}^{\infty} f_n$ *to* F] *is uniform inside D*, or that $\sum_{n=0}^{\infty} f_n$ *converges to F uniformly inside D*.

Notice that uniform convergence *on D* implies uniform convergence *inside D* but not vice versa, as the following example shows:

Example 11.1. Set $D := \Delta$ and $f_n(z) := z^n$, $n = 0, 1, \ldots$. The sequence $\{f_n\}$ converges to 0 uniformly *inside D*, but the convergence is not uniform *on D* (check!).

We will now prove that the class $H(D)$ of functions holomorphic on a domain D is closed with respect to uniform convergence inside D.

Theorem 11.7. *Let a sequence $\{f_n\}$ of complex-valued functions defined on a domain $D \subset \mathbb{C}$ converge to a function f uniformly inside D. Assume that $f_n \in H(D)$, $n = 0, 1, \ldots$. Then $f \in H(D)$. Furthermore, for every $k \in \mathbb{N}$ the sequence $\left\{ f_n^{(k)} \right\}$ converges to $f^{(k)}$ uniformly inside D.*

Proof. To prove that $f \in H(D)$, we use Theorem 11.1. First of all, as $f_n \in C(D)$, the function f is continuous on every compact subset of D, hence continuous on D (explain!). Next, fix a triangle $T \subset D$. We need to show that $\int_{\partial T} f(z)dz = 0$. Notice that by Lemma 7.2 we have $\int_{\partial T} f_n dz = 0$ for all n. Therefore, the fact that $f \in H(D)$ follows from the lemma stated below.

Lemma 11.1. *Let γ be a piecewise C^1-path in a domain $G \subset \mathbb{C}$ and $\{g_n\}$ a sequence of functions in $C(G)$. Assume that $\{g_n\}$ converges to a function g uniformly inside G (note that that $g \in C(G)$). Then $\int_\gamma g_n dz \to \int_\gamma g dz$.*

Proof. One has

$$\left| \int_\gamma g_n dz - \int_\gamma g dz \right| = \left| \int_\gamma (g_n - g)dz \right| \leq \max_{z \in \gamma([0,1])} |g_n(z) - g(z)| \, |\gamma|.$$

Since $\{g_n\}$ converges to g uniformly on the compact subset $\gamma([0,1]) \subset G$, we have

$$\max_{z \in \gamma([0,1])} |g_n(z) - g(z)| \to 0 \text{ as } n \to \infty,$$

and the lemma follows. $\square$

Further, let $k \in \mathbb{N}$ and show that $\left\{ f_n^{(k)} \right\}$ converges to $f^{(k)}$ uniformly inside D. Fix $a \in D$, $\Delta(a, r) \subset D$ and choose $0 < \rho < r$. Then by Theorem 10.4 and Lemma 11.1, for every $z \in \Delta(a, \rho)$ one has

$$f^{(k)}(z) = \frac{k!}{2\pi i} \int_{|\zeta - a| = \rho} \frac{f(\zeta)}{(\zeta - z)^{k+1}} d\zeta =$$

$$\lim_{n \to \infty} \frac{k!}{2\pi i} \int_{|\zeta - a| = \rho} \frac{f_n(\zeta)}{(\zeta - z)^{k+1}} d\zeta = \lim_{n \to \infty} f_n^{(k)}(z).$$

From the above argument it is also easy to see that the convergence of $\left\{ f_n^{(k)} \right\}$ to $f^{(k)}$ is uniform on $\Delta(a, \rho/2)$ (provide details!).

Thus, for every point $a \in D$ there exists a disk centred at a and contained in D on which $\left\{ f_n^{(k)} \right\}$ converges to $f^{(k)}$ uniformly. This proves that $\left\{ f_n^{(k)} \right\}$ converges to $f^{(k)}$ uniformly inside D (explain!). $\square$

Lemma 11.1 utilised in our proof of Theorem 11.7 can be of course re-stated for series of functions, which will be useful for us later on.

Corollary 11.5. *Suppose that γ is a piecewise C^1-path in a domain $G \subset \mathbb{C}$ and $\{g_n\}$ a sequence of functions in $C(G)$. Assume that the series $\sum_{n=0}^\infty g_n$ converges to a function g uniformly inside G. Then $\sum_{n=0}^\infty \int_\gamma g_n dz = \int_\gamma g dz$.*

Now, fix a domain $D \subset \mathbb{C}$ and consider the $\mathbb{C}$-vector space $H(D)$. It turns out that $H(D)$ can be turned into a metric space so that the metric space convergence coincides with uniform convergence inside D. Indeed, first of all, let us construct an exhaustion of D by compact subsets as follows:

$$\mathbf{K}_n := \left\{ z \in D : \text{dist}(z, \partial D) \geq \frac{1}{n}, |z| \leq n \right\}, \quad n \in \mathbb{N}. \tag{11.1}$$

Clearly, $D = \cup_{n=1}^{\infty} \mathbf{K}_n$. For every $n \in \mathbb{N}$, we define a semi-norm on $H(D)$ as (see Exercise 11.7)

$$\|f\|_{\mathbf{K}_n} := \max_{z \in \mathbf{K}_n} |f(z)|. \tag{11.2}$$

Now, for any $f, g \in H(D)$ set

$$d(f, g) := \sum_{n=1}^{\infty} \frac{\|f - g\|_{\mathbf{K}_n}}{2^n (1 + \|f - g\|_{\mathbf{K}_n})}. \tag{11.3}$$

Proposition 11.1. *The function d on $H(D) \times H(D)$ is a distance function on $H(D)$, so it turns $H(D)$ into a metric space. Furthermore, a sequence $\{f_n\}$ in $H(D)$ converges with respect to d to a function $f \in H(D)$ if and only if $\{f_n\}$ converges to f uniformly inside D.*

Proof. Homework. (Hint: use Proposition 7.2.) □

For a more general statement we refer the reader to Exercise 11.7.
    Next, we will consider series of special form.

Definition 11.7. *A power series is a series of the form $\sum_{n=0}^{\infty} c_n (z - a)^n$, where $c_n \in \mathbb{C}$ are called the coefficients of the series and $a \in \mathbb{C}$ is called the centre of the series.*

We are interested to know on what sets a power series converges.

Theorem 11.8. (Abel's Theorem) *If a power series $\sum_{n=0}^{\infty} c_n (z - a)^n$ converges at a point z_0, then it converges uniformly inside the disk $\Delta(a, |z_0 - a|)$ and the convergence is absolute on this disk.*

Proof. Assume that $z_0 \neq a$ and for any $z \in \mathbb{C}$ and any n write

$$|c_n (z - a)^n| = |c_n (z_0 - a)^n| \left| \frac{z - a}{z_0 - a} \right|^n.$$

Since the series $\sum_{n=0}^{\infty} c_n (z_0 - a)^n$ converges, one has $|c_n (z_0 - a)^n| \to 0$ as $n \to \infty$. In particular, there exists $M > 0$ such that $|c_n (z_0 - a)^n| \leq M$ for all n. Then

$$|c_n (z - a)^n| \leq M q(z)^n, \quad \text{where } q(z) := \left| \frac{z - a}{z_0 - a} \right|.$$

Let $0 < r < |z_0 - a|$. For all $z \in \overline{\Delta(a, r)}$ we have

$$q(z) \leq \frac{r}{|z_0 - a|} < 1.$$

Recall next that the series

$$\sum_{n=0}^{\infty} t^n, \text{ with } t \in \mathbb{R},$$

converges if and only if $-1 < t < 1$; in fact its sum is $1/(1-t)$. Setting $t = r/|z_0 - a|$, we see that $|c_n(z-a)^n| \le Mt^n$ for all $z \in \overline{\Delta(a,r)}$ and all n. Now, Corollary 11.4 implies that the series $\sum_{n=0}^{\infty} c_n(z-a)^n$ converges absolutely and uniformly on $\overline{\Delta(a,r)}$. As any compact subset of the disk $\Delta(a,|z_0 - a|)$ lies in $\overline{\Delta(a,r)}$ for some $0 < r < |z_0 - a|$, the theorem follows.   □

Corollary 11.6. *For any power series $\sum_{n=0}^{\infty} c_n(z-a)^n$ there exist the largest open disk centred at a, say $\Delta(a,R)$, on which the series converges (here $0 \le R \le \infty$). The convergence is absolute on $\Delta(a,R)$, uniform inside $\Delta(a,R)$, and the sum of the series is holomorphic on $\Delta(a,R)$.*

Proof. The last statement of the corollary is a consequence of Theorem 11.7.   □

Definition 11.8. For a given power series, the open disk supplied by Corollary 11.6 (which may be empty!) is called *the disk of convergence* of the series. The radius $0 \le R \le \infty$ of this disk is called *the radius of convergence* of the series.

We will now state a theorem that allows one to calculate the radius of convergence of a series via its coefficients. It will be proved in the next lecture.

Theorem 11.9. *The radius of convergence R of a power series $\sum_{n=0}^{\infty} c_n(z-a)^n$ is found from the formula*

$$R = \frac{1}{\limsup_{n \to \infty} \sqrt[n]{|c_n|}}. \tag{11.4}$$

Exercises

11.1. Let f be $\mathbb{R}$-differentiable on a domain $D \subset \mathbb{C}$ and suppose that for every $a \in D$ and every $0 < r < \text{dist}(a, \partial D)$ we have

$$\int_{|z-a|=r} f(z) dz = 0.$$

Prove that $f \in H(D)$. Can you obtain this result assuming only that $f \in C(D)$? (Hint: use Exercise 6.11.)

11.2. Let Γ be an open arc in the unit circle S^1. Suppose we have functions $f \in H(\Delta) \cap C(\overline{\Delta})$ and $g \in H(\mathbb{C} \setminus \overline{\Delta}) \cap C(\mathbb{C} \setminus \Delta)$ such that $f(z) = g(z)$ for all $z \in \Gamma$. Set

$$h(z) := \begin{cases} f(z) \text{ if } z \in \Delta \cup \Gamma, \\ g(z) \text{ if } z \in \mathbb{C} \setminus \overline{\Delta}. \end{cases}$$

Prove that $h \in H(\Delta \cup \Gamma \cup (\mathbb{C} \setminus \overline{\Delta}))$. (Hint: use Theorem 11.1.)

11.3. Give an example of a convergent series of complex numbers $\sum_{n=0}^{\infty} z_n$ such that the series $\sum_{n=0}^{\infty} z_n^3$ diverges.

11.4. Investigate the following series of complex numbers for convergence:

$$\text{(i)} \ \sum_{n=1}^{\infty} \frac{1}{n+i},$$

$$\text{(ii)} \ \sum_{n=1}^{\infty} \frac{in}{n+i}.$$

Prove your conclusions.

11.5. Suppose that a series $\sum_{n=0}^{\infty} z_n$ of complex numbers converges and there exists $0 < \alpha < \pi/2$ such that

$$|\widetilde{\arg} z_n| \leq \alpha \ \forall n,$$

where the function $\widetilde{\arg}$ was introduced in Example 7.1. Prove that the series converges absolutely.

11.6. Prove that the series

$$\sum_{n=0}^{\infty} n e^{inz}$$

converges uniformly inside the upper half-plane H and diverges at every point of the lower half-plane $\{z \in \mathbb{C} : \operatorname{Im} z < 0\}$.

11.7. Let V be a vector space over $\mathbb{R}$ or $\mathbb{C}$. Recall that a *semi-norm* on V is a function $p : V \to \mathbb{R}$ that satisfies all the conditions required for it to be a norm except the implication $p(v) = 0 \Rightarrow v = 0$, with $v \in V$. For a semi-norm p on V, an element $v \in V$ and a number $r > 0$, set

$$B_p(v,r) := \{w \in V : p(v-w) < r\}.$$

If $\{p_\alpha\}_{\alpha \in A}$ is a family of semi-norms on V, consider the topology on V with base consisting of all finite intersections $B_{p_{\alpha_1}}(v_1, r_1) \cap \cdots \cap B_{p_{\alpha_k}}(v_k, r_k)$.

(i) Prove that V with the above topology becomes a *topological vector space*, i.e., the operations of addition of vectors and multiplication of vectors by scalars are continuous maps.

(ii) Let $A = \mathbb{N}$. Assuming that the set $\{v \in V : p_n(v) = 0 \ \forall n \in \mathbb{N}\}$ contains only the zero element, prove that the function

$$d(v,w) := \sum_{n=1}^{\infty} \frac{p_n(v-w)}{2^n(1+p_n(v-w))}$$

on $V \times V$ is a metric on V that induces the topology of V.

11.8. Find the disks of convergence of the following power series:

$$(i) \sum_{n=k}^{\infty} \binom{n}{k} z^n \ \forall k \in \mathbb{N},$$

$$(ii) \sum_{n=1}^{\infty} z^{n!},$$

$$(iii) \sum_{n=0}^{\infty} n^2 (2z - i)^n,$$

$$(iv) \sum_{n=1}^{\infty} \frac{n!}{n^n} (z - 1)^n,$$

$$(v) \sum_{n=1}^{\infty} \frac{e^n}{n!} (z - 1)^n.$$

(Hint: for Parts (iv) and (v) use Stirling's approximation.)

11.9. Let $f : \mathbb{R} \to \mathbb{C}$ be a function represented as the sum of a power series in x centred at 0, i.e., there exist $c_n \in \mathbb{C}$ such that the series $\sum_{n=0}^{\infty} c_n z^n$ converges on $\mathbb{R}$ and

$$f(x) = \sum_{n=0}^{\infty} c_n x^n \ \forall x \in \mathbb{R}.$$

Prove that one can find an entire function F satisfying $F(x) = f(x)$ for all $x \in \mathbb{R}$.

11.10. Construct a function $f \in H(\Delta)$ such that f cannot be holomorphically extended to any domain larger than Δ. (Hint: try considering the power series from Part (ii) of Exercise 11.8.)

Lecture 12

Proof of Theorem 11.9. Power Series (Continued). Taylor Series. Local Power Series Expansion of a Holomorphic Function. Cauchy's Inequalities. The Uniqueness Theorem

Clearly, a power series centred at a converges on its disk of convergence $\Delta(a,R)$ and diverges at every point z satisfying $|z - a| > R$. For $R \neq 0, \infty$ it is therefore natural to investigate convergence properties of the series at the points of the circle $\{z \in \mathbb{C} : |z - a| = R\}$, which is the boundary $\partial \Delta(a,R)$ of the disk of convergence. We will now look at some examples showing that the series may: (i) diverge at every point of $\partial \Delta(a,R)$, (ii) converge on $\partial \Delta(a,R)$, (iii) converge at some points of $\partial \Delta(a,R)$ while diverging at some other points of $\partial \Delta(a,R)$.

Example 12.1. Consider *the geometric series*

$$\sum_{n=0}^{\infty} z^n,$$

for which the centre is 0 and all the coefficients are equal to 1. By formula (11.4), its disk of convergence is Δ. For any z in the unit circle $S^1 = \partial \Delta$ we have $|z|^n = 1$, thus the terms of the series do not tend to zero as $n \to \infty$, so the series diverges at every point of S^1.

One can compute the sum of the geometric series explicitly. Indeed, for any $z \in \mathbb{C}$, $z \neq 1$, the sequence of its partial sums can be found by induction:

$$F_k(z) := \sum_{n=0}^{k} z^n = \frac{1 - z^{k+1}}{1 - z}$$

(check!). Therefore, when $|z| < 1$ we have

$$F_k(z) \to \frac{1}{1 - z} \text{ as } k \to \infty.$$

Thus, on Δ one has

$$\sum_{n=0}^{\infty} z^n = \frac{1}{1 - z}.$$

Notice that the sum of the geometric series extends to a function holomorphic on $\mathbb{C} \setminus \{1\}$. The holomorphic extension of the sum is defined on $S^1 \setminus \{1\}$ despite the fact that the geometric series diverges at every point of S^1.

Example 12.2. Consider the series

$$\sum_{n=1}^{\infty} \frac{z^n}{n^2}$$

with centre 0, $c_0 = 0$ and $c_n = 1/n^2$ if $n \in \mathbb{N}$. It then follows from formula (11.4) that the disk of convergence is again Δ (check!). Furthermore, for $|z| \leq 1$ we have

$$\left| \frac{z^n}{n^2} \right| \leq \frac{1}{n^2},$$

which by Corollary 11.4 implies that the series converges absolutely and uniformly on $\overline{\Delta}$, in particular on S^1.

Example 12.3. Consider the series

$$\sum_{n=1}^{\infty} \frac{z^n}{n},$$

with centre 0, $c_0 = 0$ and $c_n = 1/n$ if $n \in \mathbb{N}$. Then formula (11.4) once again yields that the disk of convergence is Δ (check!). Now, notice that for $z = -1$ the series becomes

$$\sum_{n=1}^{\infty} \frac{(-1)^n}{n},$$

which converges (why?), whereas for $z = 1$ it becomes *the harmonic series*

$$\sum_{n=1}^{\infty} \frac{1}{n},$$

which diverges (cf. Exercise 12.1).

We will now prove Theorem 11.9. Before proceeding, recall the following:

Definition 12.1. For a sequence $\{a_n\}$ of real numbers *the limit superior of $\{a_n\}$* is defined to be

$$\limsup_{n \to \infty} a_n := \lim_{n \to \infty} \left(\sup_{m \geq n} a_m \right) = \inf_{n \geq 0} \sup_{m \geq n} a_m \in \mathbb{R} \cup \{-\infty\} \cup \{\infty\}.$$

Clearly, $\limsup_{n \to \infty} a_n$ always exists.

Proposition 12.1. *An element $A \in \mathbb{R} \cup \{\infty\}$ is equal to $\limsup_{n \to \infty} a_n$ if and only if one has*

(1) *for every $\varepsilon > 0$ there exists $N \in \mathbb{N}$ such that $a_n \leq A + \varepsilon$ for all $n \geq N$, and*

(2) *there exists a subsequence of $\{a_n\}$ converging to A.*

Also, $\limsup_{n\to\infty} a_n = -\infty$ *if and only if* $\lim_{n\to\infty}\{a_n\} = -\infty$.

Proof. Homework. □

Proof (Theorem 11.9). Set

$$A := \frac{1}{\limsup_{n\to\infty} \sqrt[n]{|c_n|}} \in \mathbb{R}_+ \cup \{\infty\}.$$

To show that $A = R$, we need to prove that the series $\sum_{n=0}^{\infty} c_n(z-a)^n$ converges on $\Delta(a,A)$ and diverges at every point z satisfying $|z-a| > A$.

Assume that $0 < A < \infty$. Fix $z_0 \in \Delta(a,A)$ and find $\varepsilon > 0$ with

$$\frac{|z_0 - a|}{A}(1+\varepsilon) < 1.$$

By condition (1) of Proposition 12.1, there exists $N \in \mathbb{N}$ such that

$$\sqrt[n]{|c_n|} \leq \frac{1+\varepsilon}{A} \quad \forall n \geq N.$$

Then

$$|c_n(z_0 - a)^n| \leq q^n \quad \forall n \geq N,$$

where

$$q := \frac{|z_0 - a|}{A}(1+\varepsilon).$$

As $0 \leq q < 1$, the series $\sum_{n=0}^{\infty} q^n$ converges, and by Corollary 11.2 we see that $\sum_{n=0}^{\infty} c_n(z-a)^n$ converges at z_0.

Next, fix z_0 with $|z_0 - a| > A$ and find $0 < \varepsilon < 1$ satisfying

$$\frac{|z_0 - a|}{A}(1-\varepsilon) \geq 1.$$

By condition (2) of Proposition 12.1, there exists a subsequence $\{c_{n_k}\}$ of the sequence $\{c_n\}$ and $K \in \mathbb{N}$ such that

$$\sqrt[n_k]{|c_{n_k}|} \geq \frac{1-\varepsilon}{A} \quad \forall k \geq K.$$

Then

$$|c_{n_k}(z_0 - a)^{n_k}| \geq \left(\frac{|z_0 - a|}{A}(1-\varepsilon)\right)^{n_k} \geq 1 \quad \forall k \geq K.$$

Therefore, the sequence $\{c_n(z_0 - a)^n\}$ does not converge to 0, which implies that the series $\sum_{n=0}^{\infty} c_n(z-a)^n$ diverges at z_0.

The proofs in the easier cases $A = 0$ and $A = \infty$ are left to the reader. □

We will now look at further properties of power series.

Proposition 12.2. (Term-by-Term Integration of Power Series) *Let f be the sum of a power series $\sum_{n=0}^{\infty} c_n(z-a)^n$ on the disk of convergence $\Delta(a,R)$, with $R > 0$, and let γ be any path in $\Delta(a,R)$. Then*

$$\int_{\gamma} f dz = \sum_{n=0}^{\infty} c_n \int_{\gamma} (z-a)^n dz = \sum_{n=0}^{\infty} \frac{c_n}{n+1} \left((\gamma(1)-a)^{n+1} - (\gamma(0)-a)^{n+1} \right).$$

Proof. As $\Delta(a,R)$ is simply-connected, the path γ is homotopic in $\Delta(a,R)$ to the segment $\mathbf{I} := [\gamma(0), \gamma(1)]$, which is a C^1-path. By Theorem 9.1 and Corollary 11.5 we then have

$$\int_{\gamma} f dz = \int_{\mathbf{I}} f dz = \sum_{n=0}^{\infty} c_n \int_{\mathbf{I}} (z-a)^n dz = \sum_{n=0}^{\infty} c_n \int_{\gamma} (z-a)^n dz,$$

and the proof is complete. □

Next, by Theorem 11.7 we have:

Proposition 12.3. (Term-by-Term Differentiation of Power Series) *Let f be the sum of a power series $\sum_{n=0}^{\infty} c_n(z-a)^n$ on the disk of convergence $\Delta(a,R)$, with $R > 0$. Then*

$$f^{(k)}(z) = \sum_{n=k}^{\infty} c_n((\zeta-a)^n)^{(k)}(z) = \sum_{n=k}^{\infty} n(n-1)\cdots(n-k+1)c_n(z-a)^{n-k} \quad (12.1)$$

for all $z \in \Delta(a,R)$ and $k \in \mathbb{N}$.

Remark 12.1. The radius of convergence of the series in the right-hand side of formula (12.1) is equal to R for all $k \in \mathbb{N}$ since $\lim_{n\to\infty, n>j}(n-j)^{\frac{1}{n}} = 1$ for all $j \in \mathbb{N}$.

Proposition 12.3 has an important consequence:

Corollary 12.1. *Let f be the sum of a power series $\sum_{n=0}^{\infty} c_n(z-a)^n$ on the disk of convergence $\Delta(a,R)$, with $R > 0$. Then*

$$c_n = \frac{f^{(n)}(a)}{n!}, \quad n = 0, 1, 2, \ldots, \text{ where } f^{(0)} := f.$$

Proof. Set $z = a$ in formula (12.1). □

Definition 12.2. Let $D \subset \mathbb{C}$ be a domain, $f \in H(D)$, and $a \in D$. *The Taylor series of f with centre a* is the power series

$$\sum_{n=0}^{\infty} \frac{f^{(n)}(a)}{n!} (z-a)^n.$$

Thus, Corollary 12.1 says that any power series with centre a coincides with the Taylor series of its sum with centre a. In particular, a power series is completely determined by its sum, i.e., it is impossible for two different power series with common centre to converge to the same function.

As we know, the sum of a power series is a function holomorphic on the corresponding disk of convergence. We will now show that, conversely, every holomorphic function can be (locally) represented as the sum of a power series.

Theorem 12.1. *Let $D \subset \mathbb{C}$ be a domain, $f \in H(D)$, $a \in D$, and $r := \mathrm{dist}(a, \partial D)$. Then on the disk $\Delta(a,r) \subset D$ the function f can be represented as the sum of a power series with centre a. Such a representation is unique and coincides with the Taylor series of f with centre a.*

Proof. The uniqueness part of the theorem follows from Corollary 12.1 as discussed above. To obtain the existence of a power series representation for f, fix a point $z \in \Delta(a,r)$ and choose $0 < \rho < r$ with $z \in \Delta(a,\rho)$. Then by Theorem 10.4 we have

$$f(z) = \frac{1}{2\pi i} \int_{|\zeta - a| = \rho} \frac{f(\zeta)}{\zeta - z} d\zeta.$$

For $|\zeta - a| > |z - a|$ write

$$\frac{1}{\zeta - z} = \frac{1}{(\zeta - a) - (z - a)} = \frac{1}{\zeta - a} \frac{1}{1 - \dfrac{z - a}{\zeta - a}} = \frac{1}{\zeta - a} \sum_{n=0}^{\infty} \left(\frac{z - a}{\zeta - a} \right)^n,$$

where in the last equality we utilised the geometric series. Note that the series

$$\frac{f(\zeta)}{\zeta - a} \sum_{n=0}^{\infty} \left(\frac{z - a}{\zeta - a} \right)^n = \sum_{n=0}^{\infty} \frac{(z - a)^n f(\zeta)}{(\zeta - a)^{n+1}}$$

converges uniformly inside the domain $\{\zeta \in \mathbb{C} : |z - a| < |\zeta - a| < r\}$. Therefore, by Corollary 11.5 we obtain

$$f(z) = \frac{1}{2\pi i} \int_{|\zeta - a| = \rho} \sum_{n=0}^{\infty} \frac{(z - a)^n f(\zeta)}{(\zeta - a)^{n+1}} d\zeta =$$

$$\sum_{n=0}^{\infty} \left(\frac{1}{2\pi i} \int_{|\zeta - a| = \rho} \frac{f(\zeta)}{(\zeta - a)^{n+1}} d\zeta \right) (z - a)^n = \sum_{n=0}^{\infty} \frac{f^{(n)}(a)}{n!} (z - a)^n,$$

where the last equality is ensured by each of Theorem 10.4 and Corollary 12.1. □

Remark 12.2. It is common to say that a function f holomorphic on a disk $\Delta(a,r)$ *expands into a power series* (*its Taylor series*) [*centred at a*] [*on $\Delta(a,r)$*] and speak about the Taylor series of f centred at a as of *the power series expansion of f* [*with centre a*] [*on $\Delta(a,r)$*].

The following upper bound on the absolute values of the coefficients of the power series expansion of a function is often useful:

Proposition 12.4. (Cauchy's Inequalities) *Let $f \in H(\Delta(a,r))$, with $0 < r \le \infty$. Then the coefficients c_n of the power series expansion of f with centre a satisfy*

$$|c_n| \le \frac{\max_{|z-a|=\rho} |f(z)|}{\rho^n}$$

for $n = 0, 1, 2, \ldots$ and every $0 < \rho < r$.

Proof. Apply Corollary 10.1. □

Theorem 12.1 has many important applications. One application is the following surprising fact:

Theorem 12.2. (The Uniqueness (or Identity) Theorem) *Let $D \subset \mathbb{C}$ be a domain and $f, g \in H(D)$. Suppose that there exists an infinite sequence $\{z_k\} \subset D$ converging to a point in D, such that $f(z_k) = g(z_k)$, $k = 0, 1, 2, \ldots$. Then $f(z) = g(z)$ for all $z \in D$.*

Proof. Define $h := f - g$. We will show that $h \equiv 0$. Set $a := \lim_{k \to \infty} z_k$. Fix a disk $\Delta(a,r) \subset D$ and expand h into a power series with centre a on $\Delta(a,r)$:

$$h(z) = c_0 + c_1(z-a) + c_2(z-a)^2 + \cdots. \tag{12.2}$$

Since $h(z_k) = 0$ for all k, we have $h(a) = 0$, i.e., $c_0 = 0$.

We shall now prove that $c_n = 0$ for all n. Assuming the opposite, suppose that c_{n_0}, $n_0 \ge 1$, is the first non-zero coefficient. Then on $\Delta(a,r)$ one has

$$
\begin{aligned}
h(z) &= c_{n_0}(z-a)^{n_0} + c_{n_0+1}(z-a)^{n_0+1} + \cdots = \\
&\quad (z-a)^{n_0}(c_{n_0} + c_{n_0+1}(z-a) + \cdots) = (z-a)^{n_0} p(z),
\end{aligned} \tag{12.3}
$$

where

$$p(z) := c_{n_0} + c_{n_0+1}(z-a) + \cdots. \tag{12.4}$$

Notice that the radius of convergence of the power series in the right-hand side of (12.4) coincides with that of the power series in the right-hand side of (12.2), thus p is defined and holomorphic on the disk $\Delta(a,r)$. Moreover, $p(a) = c_{n_0} \ne 0$, hence p does not vanish at any point of a neighbourhood U of a, with $U \subset \Delta(a,r)$. It then follows from (12.3) that the only point of U where h vanishes is a, which contradicts the fact that h is zero on the infinite sequence $\{z_k\}$ converging to a. Thus, we have proved that $c_n = 0$ for all n, therefore $h \equiv 0$ on $\Delta(a,r)$. Hence, $h^{(j)}(z) = 0$ for $j = 0, 1, \ldots$ and all $z \in \Delta(a,r)$.

Let

$$S := \{z \in D : h^{(j)}(z) = 0, \ j = 0, 1, \ldots\}.$$

The subset S is non-empty as it contains $\Delta(a,r)$. Furthermore, since each of the functions $h^{(j)}$ is continuous on D, the set S is closed in D (explain!). We will now show that S is also open in D. Indeed, fix $b \in S$ and consider a disk $\Delta(b,\rho) \subset D$. Expanding h into its Taylor series centred at b immediately yields that $h \equiv 0$ on

$\Delta(b,\rho)$, which proves that $\Delta(b,\rho)$ lies in S; hence, S is indeed open. Thus, S is a non-empty open and closed subset of D, therefore by the connectedness of D we conclude that $S = D$, so $h \equiv 0$ as required. $\square$

Notice that if the sequence $\{z_k\}$ in the statement of Theorem 12.2 converges to a *boundary point* of D, the theorem no longer holds as the following example shows:

Example 12.4. Let $D := \Delta(1,1)$ and

$$f(z) := \sin\left(\frac{1}{z}\right).$$

Clearly, f is holomorphic on D and vanishes at all points of the sequence $\left\{\dfrac{1}{\pi n}\right\}_{n\in\mathbb{N}}$, which converges to $0 \in \partial D$. However, $f \not\equiv 0$.

Exercises

12.1. Prove that the power series from Example 12.3 converges on $S^1 \setminus \{1\}$. (Hint: use Lagrange's trigonometric identities and Dirichlet's convergence test.)

12.2. Prove that the power series

$$f(z) := \sum_{n=1}^{\infty} \frac{z^n}{n^{\ln n}}$$

converges uniformly on $\overline{\Delta}$. Is this true for the power series expansion of $f^{(k)}$ with centre 0 for all $k \in \mathbb{N}$?

12.3. Find an example of a power series $\sum_{n=0}^{\infty} c_n z^n$ with radius of convergence equal to 1 such that the function

$$f : \mathbb{R} \to \mathbb{C}, \quad x \mapsto \sum_{n=0}^{\infty} c_n e^{2\pi i n x},$$

belongs to $C^{\infty}(\mathbb{R})$. Can you construct an example with the additional property that the sum of the series cannot be holomorphically extended to any domain larger than Δ? (Hint: try considering the power series $\sum_{n=0}^{\infty} 2^{-n^2} z^{2^n}$.)

12.4. Let $f \in H(\Delta)$. Define

$$g(z) := \begin{cases} \dfrac{f(z) - f(0)}{z} & \text{if } z \in \Delta \setminus \{0\}, \\ f'(0) & \text{if } z = 0. \end{cases}$$

Prove that $g \in H(\Delta)$.

12.5. Consider the domain $D := \{z \in \mathbb{C} : -1 < x < 1, \ -1 < y < 1\}$ and let $f \in H(D)$. Suppose that f cannot be holomorphically extended to any domain larger than D. Find the radii of convergence of the Taylor series of f with centres 0, $1/2 + i/2$, and $i/3$.

12.6. Let the radius of convergence R of a power series $\sum_{n=0}^{\infty} c_n(z-a)^n$ satisfy $0 < R < \infty$. Prove that for every $\varepsilon \in (0, R)$ there exists $M > 0$ such that

$$|c_n| \le \frac{M}{(R-\varepsilon)^n}, \quad n = 0, 1, 2, \ldots.$$

12.7. Let $f \in H(\Delta)$ and

$$|f'(z)| \le \frac{1}{1 - |z|} \quad \forall z \in \Delta.$$

Prove that the coefficients c_n of the power series expansion of f with centre 0 satisfy $|c_n| < e$ for all $n \in \mathbb{N}$.

12.8. Let f be an entire function. Prove that for every integer $n \ge 2$ and every $r > 0$ one has

$$\int_{|z|=r} \frac{1}{z^n} \cdot \overline{f(z)} \, dz = 0.$$

12.9. Define a function on $\mathbb{R}$ as follows:

$$f(x) := \begin{cases} e^{-\frac{1}{x^2}} & \text{if } x \ne 0, \\ 0 & \text{if } x = 0. \end{cases}$$

Prove that $f \in C^{\infty}(\mathbb{R})$ and that f cannot be represented as the sum of a power series in x with any centre on any interval containing 0.

12.10. Find the power series expansions of e^z, $\cos z$, $\sin z$ with centres $\pi/2$ and π.

12.11. Let $f \in H(\Delta) \cap C(\overline{\Delta})$. Prove that for every $\varepsilon > 0$ there exists a polynomial $P(z)$ in z such that

$$\max_{z \in \overline{\Delta}} |f(z) - P(z)| \le \varepsilon.$$

12.12. Does there exist a function $f \in H(\Delta_2)$ satisfying the condition

$$f\left(\frac{1}{n}\right) = \frac{(-1)^n}{n} \quad \forall n \in \mathbb{N}?$$

Prove your conclusion.

12.13. Does there exist a function $f \in H(\Delta_2)$ satisfying the condition

$$f\left(\frac{1}{n}\right) = e^{-n} \quad \forall n \in \mathbb{N}?$$

Prove your conclusion. (Hint: study f near the origin in the spirit of the proof of Theorem 12.2.)

12.14. Let $f \in H(\Delta)$ and $f(z) = f(z/2)$ for all $z \in \Delta$. Prove that $f \equiv \text{const}$.

12.15. Let $f \in H(\Delta) \cap C(\overline{\Delta})$ and let Γ be an open arc in the unit circle S^1. Assume that $f \equiv \text{const}$ on Γ. Prove that $f \equiv \text{const}$. (Hint: use Exercise 11.2.)

12.16. Suppose that f is an entire function and $f(x) = x^3$ for all $x \in \mathbb{R}$. Find $f(-i)$.

12.17. Let $f \in H(\mathbb{C} \setminus \mathbb{R}_-)$. Assume that $f(x) = x^x$ for all $x > 0$. Find $f(i)$.

12.18. Prove that if the compact subset $\mathbf{K}_n$ in formula (11.2) is infinite, then the semi-norm defined by this formula is in fact a norm on $H(D)$.

Lecture 13

Liouville's Theorem. Laurent Series. Annulus of Convergence. Laurent Series Expansion of a Function Holomorphic on an Annulus. Cauchy's Inequalities. Isolated Singularities of Holomorphic Functions

Another application of Theorem 12.1 is the following important fact:

Theorem 13.1. (Liouville's Theorem) *Let f be an entire function (i.e., $f \in H(\mathbb{C})$). Suppose that there exists $M > 0$ such that $|f(z)| \le M$ for all $z \in \mathbb{C}$. Then $f \equiv const.$*

In other words, an entire function with bounded absolute value is constant.

Proof. Expand f into a power series centred, say, at the origin on $\mathbb{C}$:

$$f(z) = \sum_{n=0}^{\infty} c_n z^n.$$

We will show that $c_n = 0$ for $n = 1, 2, \ldots$. Indeed, by Proposition 12.4 we have

$$|c_n| \le \frac{\max_{|z|=\rho} |f(z)|}{\rho^n} \le \frac{M}{\rho^n} \ \forall \rho > 0, \forall n.$$

Therefore, letting $\rho \to \infty$, we see that $c_n = 0$ for all $n > 0$. Hence, $f \equiv c_0$. $\square$

We shall now look at objects more general than series of functions. Let $\{f_n\}_{n \in \mathbb{Z}}$ be a collection of functions enumerated by the elements of $\mathbb{Z}$ and defined on a subset $S \subset \mathbb{C}$. Consider the formal expression

$$\sum_{n=-\infty}^{\infty} f_n. \tag{13.1}$$

It is not a series in the sense discussed in Lecture 11, so to give it a meaning we look at the two associated expressions

$$\sum_{n=0}^{\infty} f_n \quad \text{and} \quad \sum_{n=-\infty}^{-1} f_n. \tag{13.2}$$

The first one is a series of functions, whereas the second one is understood as the series $\sum_{n=1}^{\infty} f_{-n}$. Using these two series, we can speak about the convergence of $\sum_{n=-\infty}^{\infty} f_n$.

Definition 13.1. For the expression $\sum_{n=-\infty}^{\infty} f_n$, convergence of any kind (at a point, on a set, absolute, uniform, uniform inside) is understood as simultaneous convergence of that kind for the two series in (13.2). For example, we say that $\sum_{n=-\infty}^{\infty} f_n$ *converges at a point* z_0 if both $\sum_{n=0}^{\infty} f_n$ and $\sum_{n=1}^{\infty} f_{-n}$ converge at z_0, and we say that $\sum_{n=-\infty}^{\infty} f_n$ *diverges at* z_0 if at least one of the series $\sum_{n=0}^{\infty} f_n$, $\sum_{n=1}^{\infty} f_{-n}$ diverges at z_0. *The sum of* $\sum_{n=-\infty}^{\infty} f_n$ *at* z_0 is obtained by adding up the sums of $\sum_{n=0}^{\infty} f_n$ and $\sum_{n=1}^{\infty} f_{-n}$ at z_0. If $\sum_{n=-\infty}^{\infty} f_n$ converges on $S \subset \mathbb{C}$, then the resulting function $F : S \to \mathbb{C}$ is called *the sum of* $\sum_{n=-\infty}^{\infty} f_n$. In this case, we say that $\sum_{n=-\infty}^{\infty} f_n$ *converges to F [on S]* and write $F = \sum_{n=-\infty}^{\infty} f_n$.

We will now look at expressions of the form (13.1) that generalise power series.

Definition 13.2. *A Laurent series* is a formal expression

$$\sum_{n=-\infty}^{\infty} c_n(z-a)^n,$$

where $c_n \in \mathbb{C}$ are its *coefficients* and $a \in \mathbb{C}$ is its *centre*.

For a Laurent series, the two series in (13.2) become

$$\sum_{n=0}^{\infty} c_n(z-a)^n \quad \text{and} \quad \sum_{n=-\infty}^{-1} c_n(z-a)^n := \sum_{n=1}^{\infty} c_{-n}(z-a)^{-n}. \qquad (13.3)$$

The first series in (13.3) is a power series with centre a, and we can find its radius of convergence R by using formula (11.4). For the second series in (13.3) we set

$$\zeta := \frac{1}{z-a}$$

and obtain the power series

$$\sum_{n=1}^{\infty} c_{-n}\zeta^n.$$

Its radius of convergence, say r, is found from the formula

$$r = \frac{1}{\limsup_{n\to\infty} \sqrt[n]{|c_{-n}|}}.$$

Now, for any $0 \le r_1 \le \infty$, $0 \le r_2 \le \infty$ let us introduce the (open) annulus

$$\Delta(a, r_1, r_2) := \{z \in \mathbb{C} : r_1 < |z-a| < r_2\},$$

centred at a, with inner radius r_1 and outer radius r_2. We then see that $\Delta(a, 1/r, R)$ is the largest (possibly empty!) annulus with centre a on which the Laurent series $\sum_{n=-\infty}^{\infty} c_n(z-a)^n$ converges.

Definition 13.3. The annulus $\Delta(a, 1/r, R)$ is called *the annulus of convergence of the Laurent series* $\sum_{n=-\infty}^{\infty} c_n(z-a)^n$.

Clearly, $\sum_{n=-\infty}^{\infty} c_n(z-a)^n$ converges absolutely on $\Delta(a,1/r,R)$ and uniformly inside $\Delta(a,1/r,R)$, with the sum being holomorphic on $\Delta(a,1/r,R)$, and diverges at every point z satisfying either $|z-a| < 1/r$ or $|z-a| > R$. Analogously to the case of power series, the Laurent series $\sum_{n=-\infty}^{\infty} c_n(z-a)^n$ may either converge on the set

$$\{z \in \mathbb{C} : |z-a| = 1/r\} \cup \{z \in \mathbb{C} : |z-a| = R\},$$

or diverge at its every point, or converge at some of its points but not on the entire set (provide examples!).

Similarly to Propositions 12.2, 12.3 we have:

Proposition 13.1. (Term-by-Term Integration of Laurent Series) *Let f be the sum of a Laurent series $\sum_{n=-\infty}^{\infty} c_n(z-a)^n$ with annulus of convergence $\Delta(a,1/r,R)$, where $R > 1/r$, and let γ be any piecewise C^1-path in $\Delta(a,1/r,R)$. Then*

$$\int_\gamma f\,dz = \sum_{n=-\infty}^{\infty} c_n \int_\gamma (z-a)^n dz = \sum_{n=-\infty}^{-2} \frac{c_n}{n+1}\left((\gamma(1)-a)^{n+1} - (\gamma(0)-a)^{n+1}\right) +$$

$$c_{-1} \int_\gamma (z-a)^{-1} dz + \sum_{n=0}^{\infty} \frac{c_n}{n+1}\left((\gamma(1)-a)^{n+1} - (\gamma(0)-a)^{n+1}\right).$$

Proof. Homework. $\square$

Proposition 13.2. (Term-by-Term Differentiation of Laurent Series) *Let f be the sum of a Laurent series $\sum_{n=-\infty}^{\infty} c_n(z-a)^n$ with annulus of convergence $\Delta(a,1/r,R)$, where $R > 1/r$. Then*

$$f^{(k)}(z) = \sum_{n=-\infty}^{\infty} c_n((\zeta-a)^n)^{(k)}(z) = \sum_{n=-\infty}^{\infty} n(n-1)\cdots(n-k+1)c_n(z-a)^{n-k} \quad (13.4)$$

for all $z \in \Delta(a,1/r,R)$ and $k \in \mathbb{N}$.

Proof. Homework. $\square$

Notice that the annulus of convergence of the Laurent series in the right-hand side of formula (13.4) coincides with $\Delta(a,1/r,R)$ (explain!).

We will now obtain a result similar to Theorem 12.1.

Theorem 13.2. *Let $f \in H(\Delta(a,r_1,r_2))$, with $0 \le r_1 < r_2 \le \infty$. Then f can be represented on $\Delta(a,r_1,r_2)$ as the sum of a Laurent series with centre a:*

$$f(z) = \sum_{n=-\infty}^{\infty} c_n(z-a)^n.$$

Such a representation is unique, and the coefficients of this unique Laurent series are given by the formula

$$c_n = \frac{1}{2\pi i} \int_{|z-a|=\rho} \frac{f(z)}{(z-a)^{n+1}} dz \;\; \forall n \in \mathbb{Z} \quad (13.5)$$

for any $r_1 < \rho < r_2$.

Remark 13.1. Note that by Theorem 9.1 the integral in the right-hand side of formula (13.5) is independent of ρ for every $n \in \mathbb{Z}$ (explain!).

Remark 13.2. If $f \in H(\Delta(a, r_2))$, then the Laurent series supplied by Theorem 13.2 coincides with the Taylor series of f with centre a, in which case formula (13.5) for $n = 0, 1, 2, \ldots$ provides an alternative way of writing the series' coefficients (cf. the end of the proof of Theorem 12.1).

Proof (Theorem 13.2). Fix $z \in \Delta(a, r_1, r_2)$ and choose real numbers ρ_1, ρ_2 satisfying $r_1 < \rho_1 < \rho_2 < r_2$ and such that $z \in \Delta(a, \rho_1, \rho_2)$. By Theorem 10.4 we have

$$f(z) = \frac{1}{2\pi i} \int_{|\zeta-a|=\rho_2} \frac{f(\zeta)}{\zeta - z} d\zeta - \frac{1}{2\pi i} \int_{|\zeta-a|=\rho_1} \frac{f(\zeta)}{\zeta - z} d\zeta. \qquad (13.6)$$

Arguing as in the proof of Theorem 12.1, we represent the first integral in the right-hand side of (13.6) as the sum of a power series:

$$\frac{1}{2\pi i} \int_{|\zeta-a|=\rho_2} \frac{f(\zeta)}{\zeta - z} d\zeta = \sum_{n=0}^{\infty} \left(\frac{1}{2\pi i} \int_{|\zeta-a|=\rho_2} \frac{f(\zeta)}{(\zeta - a)^{n+1}} d\zeta \right) (z - a)^n =$$

$$\sum_{n=0}^{\infty} \left(\frac{1}{2\pi i} \int_{|\zeta-a|=\rho} \frac{f(\zeta)}{(\zeta - a)^{n+1}} d\zeta \right) (z - a)^n$$

for any $r_1 < \rho < r_2$, where in the last equality we appealed to Theorem 9.1.

Let us now focus on the second integral in the right-hand side of (13.6). For $|\zeta - a| < |z - a|$ write

$$\frac{1}{\zeta - z} = \frac{1}{(\zeta - a) - (z - a)} = -\frac{1}{z - a} \frac{1}{1 - \dfrac{\zeta - a}{z - a}} = -\frac{1}{z - a} \sum_{n=0}^{\infty} \left(\frac{\zeta - a}{z - a} \right)^n.$$

Note that the series

$$-\frac{f(\zeta)}{z - a} \sum_{n=0}^{\infty} \left(\frac{\zeta - a}{z - a} \right)^n = -\sum_{n=0}^{\infty} \frac{(\zeta - a)^n f(\zeta)}{(z - a)^{n+1}}$$

converges uniformly inside the domain $\{\zeta \in \mathbb{C} : r_1 < |\zeta - a| < |z - a|\}$. Therefore, by Corollary 11.5 we obtain

$$-\frac{1}{2\pi i} \int_{|\zeta-a|=\rho_1} \frac{f(\zeta)}{\zeta - z} d\zeta = \frac{1}{2\pi i} \int_{|\zeta-a|=\rho_1} \sum_{n=0}^{\infty} \frac{(\zeta - a)^n f(\zeta)}{(z - a)^{n+1}} d\zeta =$$

$$\sum_{n=0}^{\infty} \left(\frac{1}{2\pi i} \int_{|\zeta-a|=\rho_1} f(\zeta)(\zeta - a)^n d\zeta \right) \frac{1}{(z - a)^{n+1}} =$$

$$\sum_{n=1}^{\infty} \left(\frac{1}{2\pi i} \int_{|\zeta-a|=\rho_1} f(\zeta)(\zeta - a)^{n-1} d\zeta \right) (z - a)^{-n} =$$

$$\sum_{n=1}^{\infty} \left(\frac{1}{2\pi i} \int_{|\zeta-a|=\rho} f(\zeta)(\zeta-a)^{n-1} d\zeta \right)(z-a)^{-n} =$$

$$\sum_{n=-\infty}^{-1} \left(\frac{1}{2\pi i} \int_{|\zeta-a|=\rho} \frac{f(\zeta)}{(\zeta-a)^{n+1}} d\zeta \right)(z-a)^{n}$$

for any $r_1 < \rho < r_2$, where in the penultimate equality we utilised Theorem 9.1.

Thus, we have shown that the function f can be represented on $\Delta(a, r_1, r_2)$ as the sum of a Laurent series, with the coefficients c_n supplied by formula (13.5). It remains to prove the uniqueness of a Laurent series representation for f.

Suppose we are given another representation for f on $\Delta(a, r_1, r_2)$ as the sum of a Laurent series with centre a:

$$f(z) = \sum_{n=-\infty}^{\infty} d_n(z-a)^n.$$

We will show that $d_n = c_n$ for all $n \in \mathbb{Z}$. Fix $k \in \mathbb{Z}$ and for $z \in \Delta(a, r_1, r_2)$ write

$$\frac{f(z)}{(z-a)^{k+1}} = \sum_{n=-\infty}^{\infty} d_n(z-a)^{n-k-1}. \tag{13.7}$$

The Laurent series in the right-hand side of (13.7) converges uniformly inside $\Delta(a, r_1, r_2)$, hence, using Corollary 11.5, for any $r_1 < \rho < r_2$ we have

$$c_k = \frac{1}{2\pi i} \int_{|z-a|=\rho} \frac{f(z)}{(z-a)^{k+1}} dz = \frac{1}{2\pi i} \int_{|z-a|=\rho} \left(\sum_{n=-\infty}^{\infty} d_n(z-a)^{n-k-1} \right) dz =$$

$$\sum_{n=-\infty}^{\infty} d_n \frac{1}{2\pi i} \int_{|z-a|=\rho} (z-a)^{n-k-1} dz.$$

Now, the lemma stated below implies that the last expression in the above formula is equal to d_k, which completes the proof.   □

Lemma 13.1. *Let* $m \in \mathbb{Z}$. *Then for any* $\rho > 0$ *we have*

$$\frac{1}{2\pi i} \int_{|z-a|=\rho} (z-a)^m dz = \begin{cases} 1 & \text{if } m = -1, \\ 0 & \text{otherwise.} \end{cases}$$

Proof. Homework.   □

Remark 13.3. It is common to say that a function f holomorphic on an annulus $\Delta(a, r_1, r_2)$ *expands into a Laurent series* [*centred at* a] [*on* $\Delta(a, r_1, r_2)$] and speak about this unique Laurent series as of *the Laurent series expansion of* f [*with centre* a] [*on* $\Delta(a, r_1, r_2)$].

We will now record a useful upper bound on the absolute values of the coefficients of the Laurent series expansion of a function, which is analogous to Proposition 12.4.

Proposition 13.3. (Cauchy's Inequalities) *Let $f \in H(\Delta(a, r_1, r_2))$, $0 \le r_1 < r_2 \le \infty$. Then the coefficients of the Laurent series expansion of f with centre a satisfy*

$$|c_n| \le \frac{\max_{|z-a|=\rho} |f(z)|}{\rho^n}$$

for every $n \in \mathbb{Z}$ and every $r_1 < \rho < r_2$.

Proof. Homework. □

Below, we use Laurent series to study objects called *isolated singularities of holomorphic functions.*

Definition 13.4. A point $a \in \mathbb{C}$ is called *an isolated singularity of a function f*, or *an isolated singular point of f*, if $f \in H(\Delta(a, 0, r))$ for some $0 < r \le \infty$. In this case, we say that *f has an isolated singularity, or an isolated singular point, at a.*

Note here that for $r > 0$ the annulus $\Delta(a, 0, r)$ is just the disk $\Delta(a, r)$ punctured at the centre, i.e., $\Delta(a, 0, r) = \Delta(a, r) \setminus \{a\}$.

Intuitively, a "singularity", or "a singular point", of a function is a point in $\mathbb{C}$ at which the function (potentially) cannot be defined to become $\mathbb{C}$-differentiable. Notice that if f is holomorphic on the entire disk $\Delta(a, r)$, the point a can be formally considered as an isolated singularity of f despite the fact that f is known to be $\mathbb{C}$-differentiable at a. Later on, such fake isolated singularities will be called *removable.*

Isolated singularities are singularities of the simplest kind as they are "surrounded" by points at which $\mathbb{C}$-differentiability takes place. Certainly, in general a singularity does not have to be isolated as the following example shows:

Example 13.1. Let

$$f(z) := \frac{1}{\sin\left(\dfrac{1}{z}\right)}.$$

Observe that 0 is not an isolated singularity of f since the denominator vanishes at the points $\dfrac{1}{\pi n}$, $n \in \mathbb{Z}$, which accumulate to the origin when $|n| \to \infty$.

Isolated singularities can be classified into three types.

Definition 13.5. Let $a \in \mathbb{C}$ be an isolated singularity of a function f. Then

(1) a is called *a removable singularity* [*of f*] if $\lim_{z \to a} f(z)$ exists in the usual sense, i.e., one can find $A \in \mathbb{C}$ such that for every $\varepsilon > 0$ there is a sufficiently small $\delta > 0$ with $|f(z) - A| < \varepsilon$ provided $0 < |z - a| < \delta$; in this case, we say that *f has a removable singularity at a*, or that *the* [*isolated*] *singularity* [*of f*] [*at a*] *is removable*;

(2) a is called *a pole [of f]* if $\lim_{z \to a} f(z)$ exists and is equal to ∞, i.e., the following holds: $\lim_{z \to a} \Pi^{-1}(f(z)) = \infty \in \overline{\mathbb{C}}$ (we write simply $\lim_{z \to a} f(z) = \infty$); in this case, we say that f *has a pole at a*, or that *the [isolated] singularity [of f] [at a] is a pole*;

(3) a is called *an essential singularity [of f]* if $\lim_{z \to a} f(z)$ does not exist in either sense; in this case, we say that f *has an essential singularity at a*, or that *the [isolated] singularity [of f] [at a] is essential*.

Remark 13.4. In Part (2) above the condition $\lim_{z \to a} f(z) = \infty$ is equivalent to the condition $\lim_{z \to a} |f(z)| = \infty$, i.e., to the condition that for every $M > 0$ there exists a sufficiently small $\delta > 0$ such that $|f(z)| > M$ provided $0 < |z - a| < \delta$ (prove!).

It is now time for some examples.

Example 13.2. Let

$$f(z) := \frac{\sin z}{z}.$$

Clearly, 0 is an isolated singularity of f. To calculate $\lim_{z \to 0} f(z)$, recall that

$$\sin z = \frac{e^{iz} - e^{-iz}}{2i}.$$

By Example 2.2, the power series expansion of e^z with centre 0 is

$$e^z = \sum_{n=0}^{\infty} \frac{1}{n!} z^n$$

(explain!). Therefore, the power series expansion of $\sin z$ with centre 0 is

$$\sin z = \sum_{n=0}^{\infty} \frac{i^n - (-i)^n}{2in!} z^n = z + \text{higher-order terms}$$

(provide details!). Then for all $z \neq 0$ we have

$$f(z) = 1 + \text{higher-order terms},$$

hence $\lim_{z \to 0} f(z) = 1$ (explain!). Thus, f has a removable singularity at 0.

Notice that we in fact managed to represent f as the sum of a power series centred at 0 with radius of convergence $R = \infty$. In particular, we observe that extending f to all of $\mathbb{C}$ by setting $f(0) := 1$ one obtains an entire function. Later on, we will see that this is a general phenomenon for removable singularities.

Example 13.3. For $k \in \mathbb{N}$, define

$$f(z) := \frac{1}{z^k}.$$

Clearly, 0 is an isolated singularity of f and $\lim_{z \to 0} = \infty$. Therefore, f has a pole at 0.

Example 13.4. Let

$$f(z) := e^{\frac{1}{z}}.$$

As above, 0 is an isolated singularity of f. This singularity is essential as $\lim_{z \to 0} f(z)$ does not exist. Indeed, if we approach 0 along the real line $\mathbb{R}$ from the left, $f(z)$ tends to 0, whereas if we approach 0 along $\mathbb{R}$ from the right, $f(z)$ tends to ∞.

Exercises

13.1. Let f be an entire function and assume that there exist two non-zero complex numbers z_1, z_2, with $z_1/z_2 \notin \mathbb{R}$, such that

$$f(z+z_1) = f(z), \quad f(z+z_2) = f(z) \ \forall z \in \mathbb{C}.$$

Prove that $f \equiv \mathrm{const}$.

13.2. Find all entire functions f such that $|f'(z)| < |f(z)|$ for all $z \in \mathbb{C}$.

13.3. Find the annulus of convergence of the Laurent series for which the associated series from (13.3) are

$$\sum_{n=0}^{\infty} 2^n z^{n^2} \quad \text{and} \quad \sum_{n=-\infty}^{-1} 2^n z^n.$$

13.4. Let $f \in H(\Delta(a, r_1, r_2))$, with $0 \le r_1 < r_2 \le \infty$. Prove that $f(z) = g(z) + h(z)$ for all $z \in \Delta(a, r_1, r_2)$, where $g \in H(\Delta(a, r_2))$ and $h \in H(\{z \in \mathbb{C} : |z - a| > r_1\})$.

13.5. Let $f \in H(\Delta(a, r_1, r_2))$, with $0 \le r_1 < r_2 \le \infty$. Can one express the coefficients of the Laurent series expansion of f with centre a via the coefficients of the Taylor series of certain functions centred at some points?

13.6. Let $f \in H(\Delta(a, r_1, r_2))$, with $0 \le r_1 < r_2 \le \infty$, and let $r_1 < \rho < r_2$. Prove that f holomorphically extends to $\Delta(a, r_2)$ if and only if

$$\int_{|z-a|=\rho} f(z)(z-a)^n dz = 0, \ n = 0, 1, 2, \ldots.$$

13.7. Let $f \in H(\Delta(a, r_1, r_2))$, with $0 \le r_1 < r_2 < \infty$, and let $r_1 < \rho < r_2$. Prove that f holomorphically extends to $\Delta(a, r_2)$ if and only if

$$\int_{|\zeta-a|=\rho} \frac{f(\zeta)}{\zeta - z} d\zeta = 0 \ \forall z \in \mathbb{C} \setminus \overline{\Delta(a, r_2)}.$$

13.8. Let $f \in H(\Delta(a, r_1, r_2))$, with $0 < r_1 < r_2 < \infty$, and assume that $|f(z)| \le M$ for all $z \in \Delta(a, r_1, r_2)$. Prove that the coefficients of the Laurent series expansion of f with centre a satisfy

$$|c_n| \leq M\left(\frac{1}{r_1^n} + \frac{1}{r_2^n}\right)$$

for every $n \in \mathbb{Z}$.

13.9. Suppose that $f \in H(\mathbb{C} \setminus \{0\})$ and

$$|f(z)| \leq \sqrt{|z|} + \frac{1}{\sqrt{|z|}} \quad \forall z \in \mathbb{C} \setminus \{0\}.$$

Prove that $f \equiv \text{const}$.

13.10. Let $f \in H(\Delta(0,0,1/2))$ and assume that

$$|f(z)| \leq \ln\frac{1}{|z|} \quad \forall z \in \Delta(0,0,1/2).$$

Prove that 0 is a removable singularity of f and, moreover, that f holomorphically extends to $\Delta_{1/2}$.

13.11. Find all isolated singularities in $\mathbb{C}$ of the following functions and attempt to determine their types:

$$\text{(i)} \ \frac{1 - \cos z}{\sin^2 z},$$

$$\text{(ii)} \ e^{\cot \frac{\pi}{z}},$$

$$\text{(iii)} \ \frac{1 + z^{10}}{z^6(z^2 + 4)},$$

$$\text{(iv)} \ \frac{z + 1}{\sin^2 \frac{\pi}{z}},$$

$$\text{(v)} \ z^2 \sin \frac{z}{z+1},$$

$$\text{(vi)} \ \frac{1}{z^2 - 1} \cos \frac{\pi z}{z+1},$$

$$\text{(vii)} \ \cot z - \frac{1}{z},$$

$$\text{(viii)} \ \sin e^z,$$

$$\text{(ix)} \ \frac{z}{\tan z},$$

$$\text{(x)} \ \frac{z}{1 - \cos z},$$

$$\text{(xi)} \ \frac{z^2 - 1}{z^3 + 1},$$

$$\text{(xii)} \ \frac{1}{e^z - 1} - \frac{1}{\sin z},$$

(xiii) $\dfrac{1}{\cos^2 z} - \dfrac{1}{\left(z - \frac{\pi}{2}\right)^2}$,

(xiv) $\dfrac{z}{e^z + 1}$.

Lecture 14
Isolated Singularities of Holomorphic Functions (Continued). Characterisation of an Isolated Singularity via the Laurent Series Expansion. Orders of Poles and Zeroes. Casorati-Weierstrass' Theorem. Isolated Singularities of Holomorphic Functions at ∞ and their Characterisation via Laurent Series Expansions

Definition 14.1. Let $a \in \mathbb{C}$ be an isolated singularity of a function f, so we have $f \in H(\Delta(a,0,r))$ for some $0 < r \leq \infty$. By Theorem 13.2, the function f expands into a (uniquely determined) Laurent series centred at a:

$$f(z) = \sum_{n=-\infty}^{\infty} c_n(z-a)^n = \sum_{n=-\infty}^{-1} c_n(z-a)^n + \sum_{n=0}^{\infty} c_n(z-a)^n.$$

In this situation, $\sum_{n=0}^{\infty} c_n(z-a)^n$ and $\sum_{n=-\infty}^{-1} c_n(z-a)^n = \sum_{n=1}^{\infty} c_{-n}(z-a)^{-n}$ are called *the regular part* and *the principal part* of the Laurent series, respectively.

It turns out that one can characterise the type of an isolated singularity of a function via the principal part of its Laurent series expansion.

Theorem 14.1. *Let $a \in \mathbb{C}$ be an isolated singularity of a function $f \in H(\Delta(a,0,r))$, with $0 < r \leq \infty$, and let $\sum_{n=-\infty}^{\infty} c_n(z-a)^n$ be the Laurent series expansion of f with centre a. Then*
(1) *a is a removable singularity if and only if the principal part of the expansion is zero, i.e., $c_n = 0$ for all $n < 0$; in this case f extends to a function holomorphic on $\Delta(a,r)$ with $f(a) := c_0 = \lim_{z \to a} f(z)$;*
(2) *a is a pole if and only if the principal part of the expansion is non-zero and finite, i.e., there exists a negative integer N such that $c_N \neq 0$ and $c_n = 0$ for all $n < N$;*
(3) *a is an essential singularity if and only if the principal part of the expansion is infinite, i.e., there exists an infinite sequence of negative integers $\{n_k\}$ such that $c_{n_k} \neq 0$ for all k.*

Proof. Part (1). The sufficiency implication is obvious, so we assume that a is removable. Then there exist $M > 0$ and $0 < r' \leq r$ with $|f(z)| \leq M$ for $z \in \Delta(a,0,r')$ (explain!). Therefore, by Proposition 13.3 we have

$$|c_n| \leq M\rho^{-n}$$

for every $n \in \mathbb{Z}$ and every $0 < \rho < r'$. Hence, by letting $\rho \to 0$ we see that $c_n = 0$ for all $n < 0$. Thus, the Laurent series expansion of f is in fact a power series, whose sum, say F, is a function holomorphic on the disk $\Delta(a,r)$. Certainly, F coincides with f on $\Delta(a,0,r)$ and $F(a) = c_0 = \lim_{z \to a} f(z)$, so the last statement of Part (1) follows.

Part (2). Again, the sufficiency implication is obvious, so we assume that a is a pole. Then there exists $0 < r' \le r$ such that $f(z) \ne 0$ for all $z \in \Delta(a,0,r')$. Consider

$$g(z) := \frac{1}{f(z)}, \quad z \in \Delta(a,0,r').$$

Clearly, $g \in H(\Delta(a,0,r'))$ and $\lim_{z \to a} g(z) = 0$. Therefore, a is a removable singularity of g. Then, by Part (1), the Laurent series expansion of g with centre a is in fact a power series and g can be extended to a function holomorphic on $\Delta(a,r')$ with $g(a) := 0$. Write g on $\Delta(a,r')$ as

$$g(z) = d_0 + d_1(z-a) + d_2(z-a)^2 + \cdots \tag{14.1}$$

and note that $d_0 = 0$. Let $d_{n_0} \ne 0$, with $n_0 \ge 1$, be the first non-zero coefficient. Then

$$\begin{aligned} g(z) = d_{n_0}(z-a)^{n_0} + d_{n_0+1}(z-a)^{n_0+1} + \cdots = \\ (z-a)^{n_0}(d_{n_0} + d_{n_0+1}(z-a) + \cdots) = (z-a)^{n_0}p(z), \end{aligned} \tag{14.2}$$

where

$$p(z) := d_{n_0} + d_{n_0+1}(z-a) + \cdots. \tag{14.3}$$

Notice that the radius of convergence of the power series in the right-hand side of (14.3) coincides with that of the power series in the right-hand side of (14.1), thus p is defined and holomorphic on the disk $\Delta(a,r')$. Moreover, $p(a) = d_{n_0} \ne 0$, hence p does not vanish at any point of $\Delta(a,r')$. Therefore, the function

$$q(z) := \frac{1}{p(z)}, \quad z \in \Delta(a,r'),$$

is holomorphic on $\Delta(a,r')$, and we write the power series expansion of q with centre a on $\Delta(a,r')$:

$$q(z) = e_0 + e_1(z-a) + e_2(z-a)^2 + \cdots, \tag{14.4}$$

where $e_0 \ne 0$.

Then from (14.2), (14.4) we have

$$\begin{aligned} f(z) = \frac{1}{(z-a)^{n_0}}\left(e_0 + e_1(z-a) + e_2(z-a)^2 + \cdots\right) = \\ \frac{e_0}{(z-a)^{n_0}} + \cdots + e_{n_0} + e_{n_0+1}(z-a) + \cdots \quad \forall z \in \Delta(a,0,r'). \end{aligned} \tag{14.5}$$

By the uniqueness of a Laurent series expansion (see Theorem 13.2), the series in formula (14.5) is *the* Laurent series expansion of f with centre a (in particular, it converges on the larger annulus $\Delta(a,0,r)$). Clearly, this series' principal part is finite. Also, as $e_0 \ne 0$, the principal part is non-zero. We have thus established Part (2).

Part (3) is a consequence of Parts (1) and (2).   □

Notice that the above proof yields the following characterisation of removable singularities:

Proposition 14.1. *Let $a \in \mathbb{C}$ be an isolated singularity of a function $f \in H(\Delta(a,0,r))$, with $0 < r \leq \infty$. Then a is removable if and only if there exist $M > 0$ and $0 < r' \leq r$ such that $|f(z)| \leq M$ for all $z \in \Delta(a,0,r')$.*

In other words, an isolated singularity if removable if and only if the modulus of the function is bounded on some punctured disk around the point.

Proof. See the proof of Part (1) of Theorem 14.1.   □

As we will observe shortly, the proof of Theorem 14.1 also yields some additional information in the case of poles.

Definition 14.2. Let a be a pole of a function f and

$$\frac{c_{-N}}{(z-a)^N} + \cdots + \frac{c_{-1}}{z-a}$$

the principal part of the Laurent series expansion of f with centre a, where $N \in \mathbb{N}$ and $c_{-N} \neq 0$. Then the number N is called *the order of the pole* [*of f*] [*at a*]. In this situation we often say that *f has a pole of order N at a* and that *a is a pole of order N* [*of f*]. A pole of order 1 is called *a simple pole*. In this case we say that *f has a simple pole at a* and that *a is a simple pole* [*of f*].

Definition 14.3. Assume that g is holomorphic on a neighbourhood of a point $a \in \mathbb{C}$ and suppose that a is a zero of g, i.e., $g(a) = 0$. Let $K \in \mathbb{N}$ be such that $g^{(j)}(a) = 0$ for $j = 1, \ldots, K-1$ and $g^{(K)}(a) \neq 0$, so that the Taylor series of g centred at a has the form

$$c_K(z-a)^K + c_{K+1}(z-a)^{K+1} + \cdots,$$

with $c_K \neq 0$. Then the number K is called *the order of the zero* [*of g*] [*at a*]. In this situation we often say that *g has a zero of order K at a* and that *a is a zero of order K* [*of g*]. A zero of order 1 is called *a simple zero*. In this case we say that *g has a simple zero at a* and that *a is a simple zero* [*of g*].

As we will now see, there is a clear relationship between poles and zeroes of holomorphic functions.

Proposition 14.2.

(1) *Suppose that a is a zero of order K of a function g, and let $0 < r \leq \infty$ be such that $g \in H(\Delta(a,r))$ and $g(z) \neq 0$ for all $z \in \Delta(a,0,r)$. Consider the function $f := 1/g$ on $\Delta(a,0,r)$. Then f has a pole of order K at a.*

(2) *Suppose that a is a pole of order N of a function f, and let $0 < r \leq \infty$ be such that $f \in H(\Delta(a,0,r))$ and $f(z) \neq 0$ for all $z \in \Delta(a,0,r)$. Consider the function*

$$g(z) := \begin{cases} \dfrac{1}{f(z)} & \text{if } z \in \Delta(a,0,r), \\[2mm] 0 & \text{if } z = a. \end{cases}$$

Then g has a zero of order N at a.

Remark 14.1. The number r in the statement of Proposition 14.2 exists in Part (1) by Theorem 12.2 and in Part (2) by the definition of a pole (explain!).

Proof (Proposition 14.2). For a proof of Part (1) of the proposition see the proof of Part (2) of Theorem 14.1. To obtain Part (2) of the proposition, let

$$f(z) = \frac{c_{-N}}{(z-a)^N} + \cdots + \frac{c_{-1}}{z-a} + \sum_{n=0}^{\infty} c_n(z-a)^n$$

be the Laurent series expansion of f with centre a. Write this expansion in the form

$$f(z) = \frac{1}{(z-a)^N}\left(c_{-N} + c_{-N+1}(z-a) + \cdots\right) = \frac{1}{(z-a)^N} q(z),$$

where

$$q(z) := c_{-N} + c_{-N+1}(z-a) + \cdots$$

does not vanish at any point of $\Delta(a,0,r)$. In fact, q extends to a function holomorphic on $\Delta(a,r)$ and having no zeroes in $\Delta(a,r)$ (recall that $c_{-N} \neq 0$). Then

$$g(z) = (z-a)^N \frac{1}{q(z)} \quad \forall z \in \Delta(a,r). \tag{14.6}$$

Expanding

$$\frac{1}{q(z)} = d_0 + d_1(z-a) + \cdots$$

and substituting this series into (14.6), we arrive at the power series expansion of g with centre a

$$g(z) = d_0(z-a)^N + d_1(z-a)^{N+1} + \cdots.$$

Noting that $d_0 \neq 0$, we see that a is a zero of order N of g. □

We will now obtain a result showing how complicated essential singularities are.

Theorem 14.2. (Casorati-Weierstrass' Theorem) *Let $a \in \mathbb{C}$ be an essential singularity of a function $f \in H(\Delta(a,0,r))$, with $0 < r \leq \infty$. Then for every $A \in \overline{\mathbb{C}}$ there exists a sequence $\{z_n\}$ in $\Delta(a,0,r)$ converging to a such that the sequence $\{f(z_n)\}$ converges to A.*

In other words, the closure of the image of $\Delta(a,0,r)$ under f is the entire Riemann sphere.

Remark 14.2. Here and below we say that a sequence of complex numbers $\{w_n\}$ *converges to* ∞ and write $\lim_{n\to\infty} w_n = \infty$ if the sequence $\Pi^{-1}(w_n)$ converges to $\infty \in \overline{\mathbb{C}}$. This condition is equivalent to the condition $\lim_{n\to\infty} |w_n| = \infty$, i.e., to the condition that for every $M > 0$ there exists $N \in \mathbb{N}$ with $|w_n| > M$ for all $n \geq N$ (prove!).

Proof (Theorem 14.2). By Proposition 14.1, the function $|f|$ cannot be bounded on any neighbourhood of a. Therefore, for every $M > 0$ there exists $z \in \Delta(a,0,r)$ arbitrarily close to a with $|f(z)| > M$. Hence one can find a sequence $\{z_n\}$ in $\Delta(a,0,r)$ converging to a with $\lim_{n\to\infty} f(z_n) = \infty$. This establishes the theorem for $A = \infty$.

Let now $A \in \mathbb{C}$. Assume that there exists $0 < r' \leq r$ such that f does not take the value A anywhere in $\Delta(a,0,r')$ (otherwise one can find a sequence $\{z_n\}$ in $\Delta(a,0,r)$ converging to a with $f(z_n) = A$). Consider

$$g(z) := \frac{1}{f(z) - A}, \quad z \in \Delta(a,0,r').$$

Clearly, $g \in H(\Delta(a,0,r'))$, hence a is an isolated singularity of g. In fact, the singularity of g at a is essential (check!). Therefore, as shown above, there exists a sequence $\{z_n\}$ in $\Delta(a,0,r')$ converging to a for which $\lim_{n\to\infty} g(z_n) = \infty$. It then follows that $\lim_{n\to\infty} f(z_n) = A$ as required. $\square$

We will now introduce the concept of an isolated singularity at ∞.

Definition 14.4. We say that ∞ *is an isolated singularity of a function f*, or an *isolated singular point of f*, if $f \in H(\Delta(0,r,\infty))$ for some $0 \leq r < \infty$. In this case, we say that f has an isolated singularity, or an isolated singular point, at ∞.

Note here that for $r > 0$ the annulus $\Delta(0,r,\infty)$ is the complement in $\mathbb{C}$ to the closed disk $\overline{\Delta}_r$ and can be viewed as "a punctured neighbourhood of ∞" because the set $\Pi^{-1}(\Delta(0,r,\infty))$ is the intersection of an open ball in $\mathbb{R}^3$ and S^2 with ∞ deleted (explain!).

Remark 14.3. Observe that if $f \in H(\Delta(0,r,\infty))$, then for every $a \in \mathbb{C}$ there exists $0 \leq r' < \infty$ such that $f \in H(\Delta(a,r',\infty))$. This fact will be of importance for us later on.

Certainly, in general the behaviour of a holomorphic function near ∞ can be worse than having just an isolated singularity at ∞ as the following example shows:

Example 14.1. Let

$$f(z) := \frac{1}{\sin z}.$$

Notice that ∞ is not an isolated singularity of f since the denominator vanishes at the points πn, $n \in \mathbb{Z}$, which accumulate to ∞ when $|n| \to \infty$.

Isolated singularities at ∞ can be classified into three types.

Definition 14.5. Let ∞ be an isolated singularity of a function f. Then

(1) ∞ is called *a removable singularity* [*of f*] if $\lim_{z\to\infty} f(z)$ exists and lies in $\mathbb{C}$, i.e., there is $A \in \mathbb{C}$, referred to as the *limit of f at* ∞, such that $\lim_{z\to\infty\in\overline{\mathbb{C}}} f(\Pi(z)) = A$; in this case, we write $\lim_{z\to\infty} f(z) = A$ and say that f *has a removable singularity at* ∞, or that *the [isolated] singularity* [*of f*] [*at* ∞] *is removable*;

(2) ∞ is called *a pole* [*of f*] if $\lim_{z\to\infty} f(z)$ exists and is equal to ∞, i.e., the following holds: $\lim_{z\to\infty\in\overline{\mathbb{C}}} \Pi^{-1}(f(\Pi(z))) = \infty \in \overline{\mathbb{C}}$ (we write simply $\lim_{z\to\infty} f(z) = \infty$); in this case, we say that f *has a pole at* ∞, or that *the [isolated] singularity* [*of f*] [*at* ∞] *is a pole*;

(3) ∞ is called *an essential singularity* [*of f*] if $\lim_{z\to\infty} f(z)$ does not exist in either sense; in this case, we say that f *has 'an essential singularity at* ∞, or that *the [isolated] singularity* [*of f*] [*at* ∞] *is essential*.

Remark 14.4. In Part (1) above the condition $\lim_{z\to\infty} f(z) = A$ is equivalent to the condition $\lim_{|z|\to\infty} f(z) = A$, i.e., to the requirement that for every $\varepsilon > 0$ there exists a sufficiently large $R > 0$ with $|f(z) - A| < \varepsilon$ provided $|z| > R$ (cf. Exercise 10.10). In Part (2) the condition $\lim_{z\to\infty} f(z) = \infty$ is equivalent to the condition $\lim_{|z|\to\infty} |f(z)| = \infty$, i.e., to the condition that for every $M > 0$ there exists a sufficiently large $R > 0$ with $|f(z)| > M$ provided $|z| > R$ (prove both equivalences!).

Remark 14.5. Let f have an isolated singularity at ∞, so $f \in H(\Delta(0,r,\infty))$ for some $0 \le r < \infty$. Consider the function

$$g(z) := f\left(\frac{1}{z}\right), \quad z \in \Delta(0,0,1/r). \tag{14.7}$$

Obviously, $g \in H(\Delta(0,0,1/r))$, hence g has an isolated singularity at 0. Furthermore, the type of the singularity of g at 0 coincides with that of the singularity of f at ∞ (check!).

We will now give some examples.

Example 14.2. For $k \in \mathbb{N}$, set

$$f(z) := \frac{1}{z^k}.$$

Clearly, ∞ is an isolated singularity of f. As $\lim_{z\to\infty} f(z) = 0$, the function f has a removable singularity at ∞.

Example 14.3. For $k \in \mathbb{N}$, let

$$f(z) := z^k.$$

Certainly, ∞ is an isolated singularity of f and $\lim_{z\to\infty} = \infty$. Therefore, f has a pole at ∞. In fact, any non-constant polynomial $P(z) = \sum_{j=0}^{K} a_j z^j$ in z has a pole at ∞ (explain!).

Example 14.4. Define

$$f(z) := e^z.$$

As above, ∞ is an isolated singularity of f. This singularity is essential as $\lim_{z\to\infty} f(z)$ does not exist. Indeed, if we approach $-\infty$ along the real line $\mathbb{R}$, the function $f(z)$ tends to 0, whereas if we approach $+\infty$ along $\mathbb{R}$, the function $f(z)$ tends to ∞.

Definition 14.6. Let ∞ be an isolated singularity of a function f. By Remark 14.3, for every $a \in \mathbb{C}$ there exists $0 \le r < \infty$ such that $f \in H(\Delta(a,r,\infty))$. Therefore, by Theorem 13.2, for every $a \in \mathbb{C}$ the function f expands into a (uniquely determined) Laurent series centred at a on an annulus with infinite outer radius:

$$f(z) = \sum_{n=-\infty}^{\infty} c_n(z-a)^n.$$

In this situation, $\sum_{n=1}^{\infty} c_n(z-a)^n$ and $\sum_{n=-\infty}^{0} c_n(z-a)^n := \sum_{n=0}^{\infty} c_{-n}(z-a)^{-n}$ are called *the principal part* and *the regular part* of the Laurent series, respectively.

Remark 14.6. Notice that, unlike in the case of isolated singularities at points of $\mathbb{C}$, for a function having an isolated singularity at ∞ there are infinitely many Laurent series expansions, one for every choice of the centre. Thus, there are infinitely many regular and principal parts of the Laurent series expansions associated to the same function. Note also that, compared to the case of isolated singularities at points of $\mathbb{C}$, the definitions of the regular and principal parts essentially interchange when one deals with isolated singularities at ∞.

It turns out that one can characterise the type of an isolated singularity of a function f at ∞ via the principal part of *any* Laurent series expansion of f.

Theorem 14.3. *Let $\infty \in \mathbb{C}$ be an isolated singularity of a function f. Fix $a \in \mathbb{C}$ and choose $0 \le r < \infty$ such that $f \in H(\Delta(a,r,\infty))$. Let, further, $\sum_{n=-\infty}^{\infty} c_n(z-a)^n$ be the Laurent series expansion of f with centre a. Then*

(1) *∞ is a removable singularity if and only if the principal part of the expansion is zero, i.e., $c_n = 0$ for all $n > 0$;*
(2) *∞ is a pole if and only if the principal part of the expansion is non-zero and finite, i.e., there exists a positive integer N such that $c_N \ne 0$ and $c_n = 0$ for all $n > N$;*
(3) *∞ is an essential singularity if and only if the principal part of the expansion is infinite, i.e., there exists an infinite sequence of positive integers $\{n_k\}$ such that $c_{n_k} \ne 0$ for all k.*

Proof. Analogously to the function defined by formula (14.7), consider

$$g(z) := f\left(\frac{1}{z} + a\right), \quad z \in \Delta(0,0,1/r).$$

Clearly, $g \in H(\Delta(0,0,1/r))$, hence g has an isolated singularity at 0. Moreover, just as in Remark 14.5, the type of the singularity of g at 0 coincides with that of the singularity of f at ∞.

Further, it is easy to see that the Laurent series expansion of g with centre 0 is

$$g(z) = \sum_{n=-\infty}^{\infty} c_{-n} z^n.$$

In particular, the principal part of this expansion is $\sum_{n=-\infty}^{-1} c_{-n} z^n = \sum_{n=1}^{\infty} c_n z^{-n}$. The theorem now follows from Theorem 14.1. $\square$

Exercises

14.1. Determine the type of every isolated singularity in $\mathbb{C}$ of each of the functions from Exercise 13.11 and in the case of poles find their orders. Also, for each of these functions, investigate whether ∞ is one of its isolated singular points, and if so find the type.

14.2. Find the principal parts of the Laurent series expansions of

(i) $\dfrac{z-1}{\sin z}$ with centre $\pi n \; \forall n \in \mathbb{Z}$,

(ii) $\cot \pi z$ with centre $n \; \forall n \in \mathbb{Z}$.

14.3. Let $f \in H(\Delta(a,0,r))$, with $0 < r \le \infty$, and suppose that

$$f(\Delta(a,0,r)) \subset \mathbb{C} \setminus \mathbb{R}_+.$$

Prove that a is a removable singularity of f. (Hint: you may find Theorem 3.2 useful.)

14.4. Let f have an isolated singularity at $a \in \mathbb{C}$ and assume

$$\lim_{z \to a} f(z)(z-a) = 0.$$

Prove that a is a removable singularity of f.

14.5. Let $D \subset \mathbb{C}$ be a domain, $z_0 \in D$, and $f \in H(D)$. Prove that for every $n \in \mathbb{N}$ there exists $f_n \in H(D)$ such that

$$f(z) = f(z_0) + \sum_{k=1}^{n-1} \frac{f^{(k)}(z_0)}{k!}(z-z_0)^k + f_n(z)(z-z_0)^n.$$

14.6. Does there exist a function $f \in H(\mathbb{C} \setminus \{0\})$ satisfying the inequality

$$|f(z)| > e^{\frac{1}{|z|}}$$

for all $z \in \mathbb{C} \setminus \{0\}$? Prove your conclusion.

14.7. Prove that the function $\tan z - z$ has only real zeroes and find their orders. (Hint: show that $\tan z = z$ implies that the vectors $(\sin 2x, \sinh 2y)$ and (x, y) are proportional.)

14.8. Find the orders of all poles of the function

$$f(z) := \frac{e^{z^2} - 1}{z \cos z - \sin z}.$$

Determine the principal part of the Laurent series expansion of f with centre 0.

14.9. Let $a \in \mathbb{C}$ be an isolated singularity of a function $f \in H(\Delta(a, 0, r))$, where $0 < r \leq \infty$. Assume that there exist $K \geq 1$ and functions $g_j \in H(\Delta(a, r))$, $j = 0, \ldots, K - 1$, such that

$$(f(z))^K + g_{K-1}(z)(f(z))^{K-1} + \cdots + g_1(z)f(z) + g_0(z) = 0 \ \forall z \in \Delta(a, 0, r).$$

Prove that a is not an essential singularity of f. Can a be a pole of f?

14.10. Prove that the following functions have isolated singularities at ∞:

$$\text{(i)} \ e^{\sin \frac{1}{z}},$$

$$\text{(ii)} \ \frac{z^{15}}{1 + z^6},$$

$$\text{(iii)} \ \sin \pi z.$$

For each function, determine the type of the singularity and find the principal part of the Laurent series expansion on an annulus with infinite outer radius centred at each of the points 0, 1, i.

14.11. Prove that if a function f has a removable singularity at ∞, then

$$\lim_{z \to \infty} f'(z) = 0.$$

Lecture 15

Isolated Singularities of Holomorphic Functions at ∞ (Continued). Orders of Poles at ∞. Casorati-Weierstrass' Theorem for an Isolated Singularity at ∞. Residues. Cauchy's Residue Theorem. Computing Residues

We continue studying isolated singularities at ∞. Analogously to Proposition 14.1 we have:

Proposition 15.1. *Let ∞ be an isolated singularity of a function $f \in H(\Delta(0, r, \infty))$, with $0 \le r < \infty$. Then ∞ is removable if and only if there exist $M > 0$ and $r \le r' < \infty$ such that $|f(z)| \le M$ for all $z \in \Delta(0, r', \infty)$.*

Proof. We only need to prove the sufficiency implication. Write the Laurent series expansion of f with centre 0:

$$f(z) = \sum_{n=-\infty}^{\infty} c_n z^n.$$

By Proposition 13.3 we have

$$|c_n| \le M\rho^{-n}$$

for every $n \in \mathbb{Z}$ and every $r' < \rho < \infty$. Hence, by letting $\rho \to \infty$ we see that $c_n = 0$ for all $n > 0$, i.e., the principal part of the expansion is zero. By Part (1) of Theorem 14.3 we then conclude that ∞ is a removable singularity of f. $\square$

We will now define the order of a pole at ∞.

Definition 15.1. Let f have a pole at ∞. By Part (2) of Theorem 14.3, the principal part of the Laurent series expansion of f with any centre is non-zero and finite. Fix $a \in \mathbb{C}$ and consider the principal part of the Laurent series expansion of f with centre a:

$$c_1(z-a) + \cdots + c_N(z-a)^N,$$

where $N \in \mathbb{N}$ and $c_N \ne 0$. Then the number N is called *the order of the pole [of f] [at ∞]*. In this situation we often say that *f has a pole of order N at ∞* and that *∞ is a pole of order N [of f]*. A pole of order 1 is called *a simple pole*. In this case we say that *f has a simple pole at ∞* and that *∞ is a simple pole [of f]*.

To justify the above definition we need to prove:

Proposition 15.2. *The number N in Definition 15.1 is independent of a.*

Proof. Let ∞ be an isolated singularity of a function f. Fix $a \in \mathbb{C}$ and let $0 \le r < \infty$ be such that $f \in H(\Delta(a,r,\infty))$. Write the Laurent series expansion of f centred at a:

$$f(z) = \sum_{n=-\infty}^{\infty} c_n(z-a)^n.$$

Assume now that ∞ is a pole of f. Then, by Part (2) of Theorem 14.3, the principal part of this expansion is non-zero and finite, i.e., we have

$$f(z) = \sum_{n=-\infty}^{0} c_n(z-a)^n + c_1(z-a) + \cdots + c_N(z-a)^N \quad \forall z \in \Delta(a,r,\infty) \qquad (15.1)$$

for some $N \in \mathbb{N}$, where $c_N \neq 0$.

Let us attempt to obtain the Laurent series expansion of f centred at 0 from (15.1). It suffices to produce the expansion on the annulus

$$\Delta(0,|a|+r,\infty) \subset \Delta(a,r,\infty)$$

where f is holomorphic (explain!). Clearly, the principal part of expansion (15.1) can be rewritten as $d_0 + d_1 z + \cdots + d_N z^N$, where $d_N = c_N$ is non-zero. Next, for $n \in \mathbb{N}$ and $z \in \Delta(0,|a|+r,\infty)$ write

$$\frac{1}{(z-a)^n} = \frac{1}{z^n}\left(\frac{1}{1-\dfrac{a}{z}}\right)^n = \frac{1}{z^n}\left(\sum_{m=0}^{\infty}\frac{a^m}{z^m}\right)^n,$$

where we used the fact that $|z| > |a|$ for $z \in \Delta(0,|a|+r,\infty)$. Utilising Theorem 11.4, we then see that the regular part of expansion (15.1) contributes only to the regular part of the Laurent series expansion of f centred at 0 (provide details!). This shows that the principal part of the latter expansion is $d_1 z + \cdots + d_N z^N$ with $d_N \neq 0$.

Hence, it follows that the integer N in (15.1) is given by the highest-order term in the principal part of the Laurent series expansion of f centred at 0, and therefore is independent of a. □

We now note that Theorem 14.2 holds for essential singularities at ∞ as well.

Theorem 15.1. (Casorati-Weierstrass' Theorem for an Essential Singularity at ∞) *Let ∞ be an essential singularity of a function $f \in H(\Delta(0,r,\infty))$, with $0 \le r < \infty$. Then for every $A \in \overline{\mathbb{C}}$ there exists a sequence $\{z_n\}$ in $\Delta(0,r,\infty)$ converging to ∞ such that the sequence $\{f(z_n)\}$ converges to A.*

Proof. The proof is completely analogous to that of Theorem 14.2, with Proposition 15.1 utilised in place of Proposition 14.1 (provide details!). □

Next, we will introduce an important concept.

Definition 15.2. Let $a \in \mathbb{C}$ be an isolated singularity of $f \in H(\Delta(a,0,r))$ for $0 < r \leq \infty$. *The residue of f at a is*

$$\operatorname{res}_a f := \frac{1}{2\pi i} \int_{|z-a|=\rho} f(z)dz, \tag{15.2}$$

where ρ is any number satisfying $0 < \rho < r$.

By Theorem 9.1, the right-hand side of (15.2) is independent of ρ, and it is clear from formula (13.5) that $\operatorname{res}_a f$ is simply the coefficient c_{-1} of the Laurent series expansion of f with centre a.

Next, we will define the residue for an isolated singularity at ∞.

Definition 15.3. Let ∞ be an isolated singularity of $f \in H(\Delta(0,r,\infty))$ for $0 \leq r < \infty$. *The residue of f at ∞ is*

$$\operatorname{res}_\infty f := -\frac{1}{2\pi i} \int_{|z|=\rho} f(z)dz, \tag{15.3}$$

where ρ is any number satisfying $r < \rho < \infty$.

Again, by Theorem 9.1 the right-hand side of (15.3) is independent of ρ, and formula (13.5) shows that $\operatorname{res}_\infty f$ is equal to $-c_{-1}$, where c_{-1} is the coefficient at $1/z$ in the Laurent series expansion of f with centre 0.

Remark 15.1. Let ∞ be an isolated singularity of $f \in H(\Delta(0,r,\infty))$ for $0 \leq r < \infty$. Fix $a \in \mathbb{C}$ and note that $\Delta(a,|a|+r,\infty) \subset \Delta(0,r,\infty)$, hence $f \in H(\Delta(a,|a|+r,\infty))$. Observe next that any circle $\{z \in \mathbb{C} : |z| = \rho\}$ with $r < \rho < \infty$ is homotopic in $\Delta(0,r,\infty)$ to any circle $\{z \in \mathbb{C} : |z-a| = \rho'\}$ with $|a|+r < \rho' < \infty$ (prove!). Therefore, by Theorem 9.1 we have

$$\operatorname{res}_\infty f = -\frac{1}{2\pi i} \int_{|z-a|=\rho'} f(z)dz \tag{15.4}$$

for any $|a|+r < \rho' < \infty$. Notice that, by formula (13.5) the right-hand side of (15.4) is equal to the number opposite to the coefficient at $1/(z-a)$ in the Laurent series expansion of f with centre a. In particular, we see that this coefficient is independent of a.

We will now state and prove an important theorem generalising Theorem 10.1. In what follows, for a compact subset S of a domain $D \subset \mathbb{C}$, we write $f \in H(\overline{D} \setminus S)$ provided there exists a domain $G \subset \mathbb{C}$ containing $\overline{D}$ such that $f \in H(G \setminus S)$.

Theorem 15.2. (Cauchy's Residue Theorem) *Let $D \subset \mathbb{C}$ be a bounded Jordan domain and $a_1, \ldots, a_n$ points in D. Assume that $f \in H(\overline{D} \setminus \{a_1, \ldots, a_n\})$. Then*

$$\int_{\partial D} f dz = 2\pi i \sum_{j=1}^{n} \operatorname{res}_{a_j} f.$$

Proof. Choose disks $\Delta(a_j,\delta)$ such that $\overline{\Delta(a_j,\delta)} \subset D$ and $\overline{\Delta(a_j,\delta)} \cap \overline{\Delta(a_k,\delta)} = \emptyset$ for $j,k = 1,\ldots,n$, $j \neq k$. Set

$$D_\delta := D \setminus \left(\cup_{j=1}^n \overline{\Delta(a_j,\delta)} \right).$$

For the domain in Fig. 9.5 the domain D_δ is shown in Fig. 15.1, where, as before, the arrows indicate the directions of the tangent vectors to paths parametrising the components of ∂D_δ.

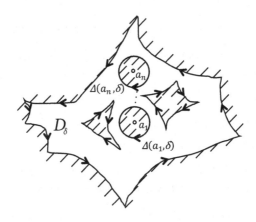

Fig. 15.1

Clearly, D_δ is a Jordan domain and $f \in H(\overline{D}_\delta)$. By Theorem 10.1 we then have

$$\int_{\partial D_\delta} f \, dz = 0.$$

On the other hand,

$$\int_{\partial D_\delta} f \, dz = \int_{\partial D} f \, dz - \sum_{j=1}^n \int_{|z-a_j|=\delta} f \, dz = \int_{\partial D} f \, dz - 2\pi i \sum_{j=1}^n \mathrm{res}_{a_j} f,$$

and the theorem follows. □

Corollary 15.1. *Let $a_1,\ldots,a_n$ be points in $\mathbb{C}$ and $f \in H(\mathbb{C} \setminus \{a_1,\ldots,a_n\})$. Then*

$$\sum_{j=1}^n \mathrm{res}_{a_j} f + \mathrm{res}_\infty f = 0.$$

Proof. Consider a disk Δ_r containing the points $a_1,\ldots,a_n$. By Theorem 15.2 we have

$$\int_{\partial \Delta_r} f \, dz = 2\pi i \sum_{j=1}^n \mathrm{res}_{a_j} f.$$

On the other hand,

$$\int_{\partial \Delta_r} f \, dz = -2\pi i \operatorname{res}_\infty f,$$

and the proof is complete.   □

Let us now discuss methods for computing residues. Apart from finding them directly from the definition, one can make a few helpful general observations as stated below.

(I). If $a \in \mathbb{C}$ is a removable singularity of a function f, then $\operatorname{res}_a f = 0$ by Part (1) of Theorem 14.1. Notice, however, that for removable singularities at ∞ the residue may not be equal to 0. For instance, we have

$$\operatorname{res}_\infty \frac{1}{z} = -1$$

(explain!).

(II). Next, we will consider the case of poles at points of $\mathbb{C}$.

Proposition 15.3. *If $a \in \overset{*}{\mathbb{C}}$ is a pole of order N of a function f, then*

$$\operatorname{res}_a f = \frac{1}{(N-1)!} \lim_{z \to a} \left((z-a)^N f(z) \right)^{(N-1)}. \tag{15.5}$$

Proof. Writing the Laurent series expansion of f with centre a

$$f(z) = \frac{c_{-N}}{(z-a)^N} + \cdots + \frac{c_{-1}}{z-a} + \sum_{n=0}^{\infty} c_n (z-a)^n,$$

we obtain

$$(z-a)^N f(z) = c_{-N} + \cdots + c_{-1}(z-a)^{N-1} + \sum_{n=0}^{\infty} c_n (z-a)^{n+N}.$$

Therefore

$$\frac{1}{(N-1)!} \lim_{z \to a} \left((z-a)^N f(z) \right)^{(N-1)} = c_{-1} = \operatorname{res}_a f$$

as required.   □

A useful formula related to (15.5) is given by:

Proposition 15.4. *Let $h, g \in H(\Delta(a, r))$ for some $0 < r \le \infty$, and assume that g has a simple zero at a (i.e., $g(a) = 0$, $g'(a) \ne 0$). Define*

$$f(z) := \frac{h(z)}{g(z)}, \quad z \in \Delta(a, 0, r'),$$

where $0 < r' \le r$ is chosen so that g does not vanish at any point of $\Delta(a, 0, r')$. Then

$$\operatorname{res}_a f := \frac{h(a)}{g'(a)}. \tag{15.6}$$

Proof. Homework. (Hint: notice that a is either a removable singularity (if $h(a) = 0$) or a simple pole (if $h(a) \neq 0$) of f and use Proposition 15.3 in the latter case.) □

Rather than applying formulas (15.5), (15.6), it is often more convenient to attempt to directly compute the coefficient c_{-1} of the relevant Laurent series expansion. We shall now illustrate this approach with examples.

Example 15.1. Set

$$f(z) := \frac{z - \frac{\pi}{2}}{\cos^2 z}$$

and find $\mathrm{res}_{\pi/2} f$. As we will see shortly, f has a simple pole at $\pi/2$. Indeed, write the power series expansion of $\cos z$ centred at $\pi/2$:

$$\cos z = -\left(z - \frac{\pi}{2}\right) + \frac{\left(z - \frac{\pi}{2}\right)^3}{6} + \cdots. \tag{15.7}$$

This yields

$$\cos^2 z = \left(z - \frac{\pi}{2}\right)^2 - \frac{\left(z - \frac{\pi}{2}\right)^4}{3} + \cdots = \left(z - \frac{\pi}{2}\right)^2 p(z),$$

where

$$p(z) := 1 - \frac{\left(z - \frac{\pi}{2}\right)^2}{3} + \cdots.$$

As p is holomorphic on a neighbourhood of $\pi/2$ and $p(\pi/2) = 1 \neq 0$, the function $1/p$ is holomorphic near $\pi/2$. Writing the power series expansion of $1/p$ centred at $\pi/2$

$$\frac{1}{p(z)} = 1 + d_1 \left(z - \frac{\pi}{2}\right) + d_2 \left(z - \frac{\pi}{2}\right)^2 + \cdots,$$

we are led to the Laurent series expansion of f with centre $\pi/2$:

$$f = \left(z - \frac{\pi}{2}\right) \frac{1}{\left(z - \frac{\pi}{2}\right)^2} \frac{1}{p(z)} = \frac{1}{z - \frac{\pi}{2}} \left(1 + d_1 \left(z - \frac{\pi}{2}\right) + d_2 \left(z - \frac{\pi}{2}\right)^2 + \cdots\right) = $$
$$\frac{1}{z - \frac{\pi}{2}} + d_1 + d_2 \left(z - \frac{\pi}{2}\right) + \cdots.$$

It then follows that f has a simple pole at $\pi/2$ and $\mathrm{res}_{\pi/2} f = 1$ (explain!).

Example 15.2. Let

$$f(z) := \frac{z - \frac{\pi}{2}}{\cos^3 z}.$$

Again, we wish to compute $\mathrm{res}_{\pi/2} f$. As we will see below, f has a pole of order 2 at $\pi/2$. Using (15.7) we obtain

$$\cos^3 z = -\left(z - \frac{\pi}{2}\right)^3 + \frac{\left(z - \frac{\pi}{2}\right)^5}{2} + \cdots = \left(z - \frac{\pi}{2}\right)^3 p(z),$$

where

$$p(z) := -1 + \frac{\left(z - \frac{\pi}{2}\right)^2}{2} + \cdots. \tag{15.8}$$

Since p is holomorphic on a neighbourhood of $\pi/2$ and $p(\pi/2) = -1 \neq 0$, the function $1/p$ is holomorphic near $\pi/2$. Writing the power series expansion of $1/p$ with centre $\pi/2$

$$\frac{1}{p(z)} = -1 + d_1\left(z - \frac{\pi}{2}\right) + d_2\left(z - \frac{\pi}{2}\right)^2 + \cdots,$$

we obtain the Laurent series expansion of f with centre $\pi/2$:

$$f = \left(z - \frac{\pi}{2}\right) \frac{1}{\left(z - \frac{\pi}{2}\right)^3} \frac{1}{p(z)} =$$

$$\frac{1}{\left(z - \frac{\pi}{2}\right)^2}\left(-1 + d_1\left(z - \frac{\pi}{2}\right) + d_2\left(z - \frac{\pi}{2}\right)^2 + \cdots\right) =$$

$$-\frac{1}{\left(z - \frac{\pi}{2}\right)^2} + \frac{d_1}{z - \frac{\pi}{2}} + d_2 + \cdots.$$

Therefore, f has a pole of order 2 at $\pi/2$ and $\operatorname{res}_{\pi/2} f = d_1$ (explain!).

Let us now determine d_1, Clearly, we have

$$d_1 = \left(\frac{1}{p}\right)'\left(\frac{\pi}{2}\right) = -\frac{p'\left(\frac{\pi}{2}\right)}{p^2\left(\frac{\pi}{2}\right)} = 0,$$

where we utilised (15.8) to conclude that $p'\left(\frac{\pi}{2}\right) = 0$. We thus see that $\operatorname{res}_{\pi/2} f = 0$.

Exercises

15.1. For each of the functions listed in Exercise 13.11, determine whether ∞ is one of its poles, and if so find the order.

15.2. Assume that a domain $D \subset \mathbb{C}$ is *symmetric about the origin*, i.e., $z \in D$ implies $-z \in D$. Let $a \in D$ and $f \in H(D \setminus \{a, -a\})$. Prove the following statements:

(i) if f is even (i.e., $f(z) = f(-z)$ for all $z \in D$, $z \neq a, -a$), then $\operatorname{res}_a f = -\operatorname{res}_{-a} f$,

(ii) if f is odd (i.e., $f(z) = -f(-z)$ for all $z \in D$, $z \neq a, -a$), then $\operatorname{res}_a f = \operatorname{res}_{-a} f$.

15.3. Let $f \in H(\Delta(a, r, \infty))$ for some $0 \leq r < \infty$ and assume that $\mathrm{res}_\infty f = 0$. Prove that f has a holomorphic primitive on $\Delta(a, r, \infty)$.

15.4. State and prove a variant of Cauchy's Residue Theorem for unbounded domains. Namely, if $D \subset \mathbb{C}$ be an unbounded Jordan domain, $a_1, \ldots, a_n \in D$, and $f \in H(\overline{D} \setminus \{a_1, \ldots, a_n\})$, what is the value of the integral

$$\int_{\partial D} f \, dz \, ?$$

Prove your conclusion.

15.5. Compute the integrals:

$$(\mathrm{i}) \int_{|z|=1} \frac{z}{z^2 + z - 1} \, dz,$$

$$(\mathrm{ii}) \int_{|z|=2} \frac{z^{100} + 2z}{z^{101} + 1} \, dz.$$

15.6. Let $D \subset \mathbb{C}$ be a bounded Jordan domain, $f \in H(\overline{D})$, and P a polynomial in z of degree $K \geq 1$ with K pairwise distinct roots $w_1, \ldots, w_K$ lying in D. Prove that the function

$$Q(z) := \frac{1}{2\pi i} \int_{\partial D} \frac{f(\zeta)}{P(\zeta)} \frac{P(\zeta) - P(z)}{\zeta - z} d\zeta, \ z \in \mathbb{C},$$

is a polynomial in z of degree not exceeding $K - 1$ and $Q(w_j) = f(w_j), j = 1, \ldots, K$. (Hint: notice that the ratio $(P(\zeta) - P(z))/(\zeta - z)$ is an entire function of ζ for every fixed z.)

15.7. For $K \geq 2$, let $w_1, \ldots, w_K$ be pairwise distinct complex numbers and P a polynomial in z with $\deg P \leq K - 2$. Set $Q(z) := (z - w_1) \cdots (z - w_K)$. Prove that

$$\sum_{j=1}^{K} \frac{P(w_j)}{Q'(w_j)} = 0.$$

(Hint: consider the integral of the ratio $P(z)/Q(z)$ over the boundary of a disk of sufficiently large radius.)

15.8. For every $n \in \mathbb{N}$, compute the residue

$$\mathrm{res}_\infty \frac{1}{z^n - 1}$$

directly from the definition and, independently, by utilising Corollary 15.1.

15.9. For every $n \in \mathbb{N}$, compute the residue

$$\mathrm{res}_1 \frac{e^z}{(z - 1)^n}$$

by appealing to formula (15.5) and, independently, by utilising Laurent series expansions as in Examples 15.1, 15.2.

15.10. Compute $\mathrm{res}_{\pi/2} f$ for the following choices of f:

$$\text{(i)} \quad \frac{z - \frac{\pi}{2}}{\cos^4 z},$$

$$\text{(ii)} \quad \frac{z - \frac{\pi}{2}}{\cos^5 z}.$$

In each case, can you find the principal part of the Laurent series expansion of f with centre $\pi/2$?

Lecture 16
Computing Residues (Continued). Computing Integrals over the Real Line Using Contour Integration. The Argument Principle

We continue discussing recipes for computing residues.

(III). For essential singularities, the only available general method to compute residues is via Laurent series expansions.

Example 16.1. Set

$$f(z) := e^{\frac{1}{z}}.$$

As we saw in Example 13.4, the origin is an essential singularity of f. Let us find $\mathrm{res}_0 f$. We know

$$e^z = \sum_{n=0}^{\infty} \frac{z^n}{n!} \quad \forall z \in \mathbb{C}.$$

Hence, for all $z \neq 0$ we have

$$f(z) = \sum_{n=0}^{\infty} \frac{1}{n!} z^{-n} = 1 + \frac{1}{z} + \frac{1}{2z^2} + \cdots,$$

which is the Laurent series expansion of f with centre 0. By Part (3) of Theorem 14.1, this proves, once again, that 0 is an essential singularity of f (explain!). Furthermore, we see that $\mathrm{res}_0 f = 1$.

(IV). To find residues at ∞, one can either use Corollary 15.1 or to determine the coefficient c_{-1} of the Laurent series expansion centred at a conveniently chosen point.

Example 16.2. Fix an integer $n \geq 2$ and let

$$f(z) := \frac{1}{z^n + 1}.$$

Clearly, ∞ is a removable singularity of f. Our goal is to compute $\mathrm{res}_\infty f$. By Corollary 15.1 one has

$$\text{res}_\infty f = - \sum_{j=1}^{n} \text{res}_{a_j} f,$$

where $a_1, \dots, a_n$ are the roots of -1 of order n. By Proposition 15.4, for every j we have

$$\text{res}_{a_j} f = \frac{1}{n a_j^{n-1}} = -\frac{a_j}{n}$$

(explain!). Therefore,

$$\text{res}_\infty f = \frac{1}{n} \sum_{j=1}^{n} a_j = 0.$$

Alternatively, one can find the Laurent series expansion of f centred at 0. We will write this expansion on the annulus $\Delta(0, 1, \infty)$ as follows:

$$f(z) = \frac{1}{z^n} \frac{1}{1 + \dfrac{1}{z^n}} = \frac{1}{z^n} \sum_{k=0}^{\infty} \frac{(-1)^k}{z^{nk}}.$$

The coefficient c_{-1} of the above series is zero, hence we see, as before, that $\text{res}_\infty f = 0$.

Finally, one can compute $\text{res}_\infty f$ directly from the definition. Indeed,

$$|\text{res}_\infty f| = \frac{1}{2\pi} \left| \int_{|z|=\rho} f(z) dz \right| \leq \max_{|z|=\rho} |f(z)| \rho$$

for any $1 < \rho < \infty$, which is a special case of Proposition 13.3. Letting $\rho \to \infty$, we clearly see that $\max_{|z|=\rho} |f(z)| \rho \to 0$. Hence, once again, we obtain that $\text{res}_\infty f = 0$.

All in all, apart from finding residues directly from the definition, the only general method applicable to isolated singularities of all kinds is to utilise Laurent series expansions.

We will now briefly touch on one significant application of residues, namely, on computing integrals arising in real analysis by complex-analytic methods (the totality of such methods is often referred to as *contour integration*). There are many types of integrals that can be calculated by using contour integration, but we only look at integrals of one kind in order to illustrate the idea behind this approach.

Consider a rational function

$$f(x) = \frac{P(x)}{Q(x)}$$

on $\mathbb{R}$, where P, Q are real-valued polynomials in x, with $\deg Q \geq \deg P + 2$ and Q everywhere non-vanishing. We are interested in finding the improper Riemann integral

$$\int_{-\infty}^{\infty} f(x) dx := \lim_{R \to \infty} \int_{-R}^{0} f(x) dx + \lim_{R \to \infty} \int_{0}^{R} f(x) dx.$$

Notice that, owing to the assumptions on P and Q, each of the above limits exists, so we can compute the integral as follows:

$$\int_{-\infty}^{\infty} f(x)dx = \lim_{R \to \infty} \int_{-R}^{R} f(x)dx. \tag{16.1}$$

Let us extend the function f to (most of) the complex plane by setting

$$F(z) := \frac{P(z)}{Q(z)},$$

where we write $P(z)$ and $Q(z)$ for the natural polynomial extensions of $P(x)$ and $Q(x)$ to $\mathbb{C}$, respectively, obtained by substituting z in place of x. As the coefficients of the polynomial $Q(z)$ are real numbers, the set of its pairwise distinct roots splits into two subsets:

$$\{a_1,\dots,a_n\} \subset H = \{z \in \mathbb{C} : \operatorname{Im} z > 0\} \text{ and } \{\bar{a}_1,\dots,\bar{a}_n\} \subset \{z \in \mathbb{C} : \operatorname{Im} z < 0\},$$

with $a_j \neq a_k$ for $j \neq k$. Clearly, $F \in H(\mathbb{C} \setminus \{a_1,\dots,a_n,\bar{a}_1,\dots,\bar{a}_n\})$.
For every $R > 0$ let

$$D_R := \Delta_R \cap H$$

and suppose that R is large enough to guarantee that $a_j \in D_R$ for $j = 1,\dots,n$ (see Fig. 16.1).

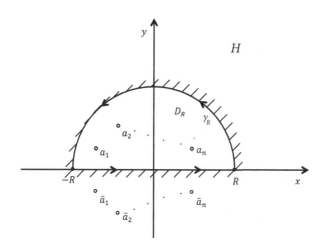

Fig. 16.1

By Theorem 15.2, we have

$$\int_{\partial D_R} F(z)dz = 2\pi i \sum_{j=1}^{n} \operatorname{res}_{a_j} F. \tag{16.2}$$

On the other hand,

$$\int_{\partial D_R} F(z)dz = \int_{-R}^{R} f(x)dx + \int_{\gamma_R} F(z)dz, \tag{16.3}$$

where $\gamma_R(t) := Re^{\pi i t}$ (explain!).

We now let $R \to \infty$ and note that the second integral in formula (16.3) tends to zero. Indeed

$$\left| \int_{\gamma_R} F(z)dz \right| \leq \pi \max_{z \in \gamma_R} |F(z)| R,$$

and $\max_{z \in \gamma_R} |F(z)| R \to 0$ by the assumption $\deg Q \geq \deg P + 2$ (explain!). Therefore, by formulas (16.1), (16.2), (16.3) we have

$$\int_{-\infty}^{\infty} f(x)dx = \lim_{R \to \infty} \int_{\partial D_R} F(z)dz = 2\pi i \sum_{j=1}^{n} \operatorname{res}_{a_j} F. \tag{16.4}$$

In words, formula (16.4) allows one to calculate the integral of a rational function over $\mathbb{R}$ via the residues of its extension to the complex plane at all pairwise distinct roots of the denominator that lie in the upper half-plane H. We will now illustrate this method for finding integrals with examples.

Example 16.3. First, let us compute

$$\int_{-\infty}^{\infty} \frac{1}{x^2 + 1}dx. \tag{16.5}$$

Here $P(x) := 1$, $Q(x) := x^2 + 1$. Therefore, $P \equiv 1$, $Q(z) = z^2 + 1$, and

$$F(z) = \frac{1}{z^2 + 1}.$$

Clearly, the roots of $Q(z)$ are $\pm i$. Thus, by formula (16.4) we have

$$\int_{-\infty}^{\infty} \frac{1}{x^2 + 1}dx = 2\pi i \operatorname{res}_i F = 2\pi i \frac{1}{2i} = \pi,$$

where we utilised Proposition 15.4 for computing $\operatorname{res}_i F$ (provide details!).

Note that integral (16.5) can be also found directly by using a primitive:

$$\int_{-\infty}^{\infty} \frac{1}{x^2 + 1}dx = \lim_{R \to \infty} \int_{-R}^{R} \frac{1}{x^2 + 1}dx = \lim_{R \to \infty} (\tan^{-1}(R) - \tan^{-1}(-R)) = \pi.$$

We will now look at a more interesting example.

Example 16.4. Let us find

$$\int_{-\infty}^{\infty} \frac{1}{x^4 + 1}dx.$$

Here $P(x) := 1$, $Q(x) := x^4 + 1$. Therefore, $P \equiv 1$, $Q(z) = z^4 + 1$, and

$$F(z) = \frac{1}{z^4 + 1}.$$

The roots of $Q(z)$ lying in H are

$$a_1 := \frac{1+i}{\sqrt{2}}, \quad a_2 := \frac{-1+i}{\sqrt{2}}.$$

Thus, by formula (16.4) we have

$$\int_{-\infty}^{\infty} \frac{1}{x^4 + 1} dx = 2\pi i (\operatorname{res}_{a_1} F + \operatorname{res}_{a_2} F) =$$

$$2\pi i \left(\frac{1}{4a_1^3} + \frac{1}{4a_2^3} \right) = 2\pi i \left(-\frac{1+i}{4\sqrt{2}} + \frac{1-i}{4\sqrt{2}} \right) = \frac{\pi}{\sqrt{2}},$$

where we used Proposition 15.4 to compute $\operatorname{res}_{a_1} F$ and $\operatorname{res}_{a_2} F$ (provide details!).

Our next goal is to obtain an important result, called the Argument Principle. We will start with the following fact:

Proposition 16.1.

(1) *Suppose that $a \in \mathbb{C}$ is a pole of order N of a function f, and let $0 < r \leq \infty$ be such that $f \in H(\Delta(a, 0, r))$ and $f(z) \neq 0$ for $z \in \Delta(a, 0, r)$. Then for the function $f'/f \in H(\Delta(a, 0, r))$ we have*

$$\operatorname{res}_a \frac{f'}{f} = -N. \tag{16.6}$$

(2) *Suppose that $a \in \mathbb{C}$ is a zero of order K of a function g, and let $0 < r \leq \infty$ be such that $g \in H(\Delta(a, r))$ and $g(z) \neq 0$ for $z \in \Delta(a, 0, r)$. Then for the function $g'/g \in H(\Delta(a, 0, r))$ we have*

$$\operatorname{res}_a \frac{g'}{g} = K. \tag{16.7}$$

Proof. To obtain Part (1), write f on $\Delta(a, 0, r)$ as

$$f(z) = \frac{q(z)}{(z-a)^N},$$

where $q \in H(\Delta(a, r))$ and $q(z) \neq 0$ for $z \in \Delta(a, r)$ (see the proof of Part (2) of Proposition 14.2). Then

$$\frac{f'}{f}(z) = \frac{\dfrac{q'(z)}{(z-a)^N} - \dfrac{Nq(z)}{(z-a)^{N+1}}}{\dfrac{q(z)}{(z-a)^N}} = \frac{q'(z)}{q(z)} - \frac{N}{z-a}.$$

Since $q'/q \in H(\Delta(a,r))$, formula (16.6) follows.

For Part (2), write g on $\Delta(a,r)$ as

$$g(z) = (z-a)^K p(z),$$

where $p \in H(\Delta(a,r))$ and $p(z) \neq 0$ for $z \in \Delta(a,r)$ (cf. the proof of Part (2) of Theorem 14.1). Then

$$\frac{g'}{g}(z) = \frac{(z-a)^K p'(z) + K(z-a)^{K-1}p(z)}{(z-a)^K p(z)} = \frac{p'(z)}{p(z)} + \frac{K}{z-a}.$$

As $p'/p \in H(\Delta(a,r))$, we obtain (16.7). $\quad\square$

We will now state an initial variant of the Argument Principle.

Theorem 16.1. (The Argument Principle, Preliminary Version) *Let $D \subset \mathbb{C}$ be a bounded Jordan domain, $a_1,\ldots,a_n$ points in D, and $f \in H(\overline{D} \setminus \{a_1,\ldots,a_n\})$. Assume that f has a pole of order N_j at a_j, $j = 1,\ldots,n$. Suppose further that f does not vanish at any point of ∂D and let $b_1,\ldots,b_k$ be the zeroes of f in D. Define K_ℓ to be the order of b_ℓ, $\ell = 1,\ldots,k$. Then*

$$\int_{\partial D} \frac{f'}{f}\,dz = 2\pi i(K - N), \tag{16.8}$$

where $K := \sum_{\ell=1}^k K_\ell$ and $N := \sum_{j=1}^n N_j$.

Remark 16.1. By the assumption $f(z) \neq 0$ for all $z \in \partial D$, Theorem 12.2 implies that the function f has only finitely many zeroes in D (explain!).

Proof (Theorem 16.1). Observe that $f'/f \in H(\overline{D} \setminus \{a_1,\ldots,a_n,b_1,\ldots,b_k\})$. By Theorem 15.2 we have

$$\int_{\partial D} \frac{f'}{f}\,dz = 2\pi i \left(\sum_{j=1}^n \operatorname{res}_{a_j} \frac{f'}{f} + \sum_{\ell=1}^k \operatorname{res}_{b_\ell} \frac{f'}{f} \right).$$

The theorem now follows from Proposition 16.1. $\quad\square$

Next, to deduce the standard variant of the Argument Principle, we will interpret Theorem 16.1 in different terms. First, we need:

Proposition 16.2. *For any closed path γ in $\mathbb{C} \setminus \{0\}$ there exists a unique integer m such that γ is homotopic in $\mathbb{C} \setminus \{0\}$ to the path $e^{2\pi i m t}$.*

Proof. Without loss of generality, we can suppose that $|\gamma(t)| = 1$ for all $t \in [0,1]$. Indeed,

$$\Gamma(t,s) := (1-s)\gamma(t) + s\frac{\gamma(t)}{|\gamma(t)|}$$

is a CP-homotopy between γ and $\gamma/|\gamma|$ in $\mathbb{C}\setminus\{0\}$. Furthermore, we can suppose that $\gamma(0) = 1$ (explain!). Under these assumptions, we have

$$\gamma(t) = e^{2\pi i\phi(t)}, \tag{16.9}$$

where $\phi : [0, 1] \to \mathbb{R}$ is a function with $\phi(0) = 0$.

We claim that one can choose ϕ to be continuous. We will only sketch a proof of this fact leaving details to the reader. Indeed, let us construct a primitive Φ of the function $1/z$ along γ by following the four-step procedure described in Lecture 7. Namely, we proceed as in Example 7.1 utilising the functions $\ln z$ and $\ln_0 z$. Then we set

$$\phi(t) := \frac{1}{2\pi i}\Phi(t), \ t \in [0, 1].$$

Clearly, ϕ is continuous, $\phi(0) = 0$, and formula (16.9) holds as

$$2\pi\phi(t) = -i\Phi(t) \in \text{Arg } \gamma(t) \ \forall t \in [0, 1]$$

by construction. Since γ is closed, $\phi(1)$ is an integer; we call it m.

Now,

$$\widetilde{\Gamma}(t, s) := e^{2\pi i((1-s)\phi(t)+smt)}$$

is a CP-homotopy between γ and $e^{2\pi imt}$ in $\mathbb{C}\setminus\{0\}$ as required.

The uniqueness of $m \in \mathbb{Z}$ for which γ is homotopic in $\mathbb{C}\setminus\{0\}$ to $e^{2\pi imt}$ follows from Example 6.1 and Theorem 9.1 (explain!). $\square$

Exercises

16.1. Find the following residues:

$$(i) \ \text{res}_1 \ ze^{\frac{1}{z-1}},$$

$$(ii) \ \text{res}_\infty \frac{\sin\frac{1}{z}}{z-1},$$

$$(iii) \ \text{res}_\infty z^2 \sin\frac{\pi}{z},$$

$$(iv) \ \text{res}_\infty z^2 \sin e^{\frac{1}{z}},$$

$$(v) \ \text{res}_0 \frac{e^{z^2}}{z^{2n+1}} \qquad \forall n \in \mathbb{N},$$

$$(vi) \ \text{res}_\infty z^n e^{\frac{a}{z}} \qquad \forall a \in \mathbb{C}, \forall n \in \mathbb{N},$$

$$(vii) \ \text{res}_0 \frac{1}{(1-e^{-z})^n} \qquad \forall n \in \mathbb{N},$$

$$\text{(viii) res}_0 \frac{z^{n-1}}{\sin^n z} \quad \forall n \in \mathbb{N}.$$

(Hint: in Part (vii) use induction on n.)

16.2. Find the residue at every isolated singularity in $\overline{\mathbb{C}}$ of each of the following functions:

$$\text{(i)} \quad \frac{1+z^8}{z^6(z+2)},$$

$$\text{(ii)} \quad \frac{1+z^{2n}}{z^n(z-a)} \quad \forall a \in \mathbb{C}, \ \forall n \in \mathbb{N},$$

$$\text{(iii)} \ \sin z \cdot \sin \frac{1}{z},$$

$$\text{(iv)} \ e^{\sin \frac{1}{z}}.$$

16.3. Find the following integrals by using contour integration:

$$\text{(i)} \ \int_{-\infty}^{\infty} \frac{x^2}{(x^2+a^2)^3}\,dx \qquad \forall a > 0,$$

$$\text{(ii)} \ \int_{-\infty}^{\infty} \frac{1}{(x^2+a^2)(x^2+b^2)^2}\,dx \ \forall a > 0, \ \forall b > 0, \ a \neq b,$$

$$\text{(iii)} \ \int_{-\infty}^{\infty} \frac{x^2}{x^4+6x^2+25}\,dx,$$

$$\text{(iv)} \ \int_{-\infty}^{\infty} \frac{2x^2-4x+5}{(x-2)^3(x^2+1)}\,dx,$$

where the last integral is understood as follows:

$$\int_{-\infty}^{\infty} \frac{2x^2-4x+5}{(x-2)^3(x^2+1)}\,dx :=$$
$$\lim_{R \to \infty} \left(\lim_{\varepsilon \to 0} \left(\int_{-R}^{2-\varepsilon} \frac{2x^2-4x+5}{(x-2)^3(x^2+1)}\,dx + \int_{2+\varepsilon}^{R} \frac{2x^2-4x+5}{(x-2)^3(x^2+1)}\,dx \right) \right).$$

(Hint: in Part (iv) allow the contour of integration to include a semi-circle centred at 2 and express the limit of the integral of the extended function over this semi-circle—as its radius tends to zero—via the residue at 2.)

16.4. Let $P(z)$, $Q(z)$ be two polynomials in z with $\deg Q > \deg P$, and $\alpha > 0$. Prove that

$$\lim_{R \to \infty} \int_{\gamma_R} \frac{P(z)}{Q(z)} e^{i\alpha z}\,dz = 0,$$

where $\gamma_R(t) := Re^{\pi i t}$.

16.5. Using Exercise 16.4, find the following limit:

$$\lim_{R \to \infty} \int_{-R}^{R} \frac{x \sin x}{x^2 + a^2} dx \ \forall a > 0.$$

16.6. Find $\mathrm{res}_0 \dfrac{f'}{f}$ for the following choices of f:

$$\text{(i) } z^3 \sin^2 z,$$

$$\text{(ii) } z^2 \cot^5 z,$$

$$\text{(iii) } \frac{z^8}{(\cos z - 1)^4}.$$

16.7. Let $D \subset \mathbb{C}$ be a bounded Jordan domain, $a_1, \ldots, a_n$ points in D, and $f \in H(\overline{D} \setminus \{a_1, \ldots, a_n\})$, $g \in H(\overline{D})$. Assume that f has a pole of order N_j at a_j, $j = 1, \ldots, n$. Suppose further that f does not vanish at any point of ∂D and let $b_1, \ldots, b_k$ be the zeroes of f in D. Define K_ℓ to be the order of b_ℓ, $\ell = 1, \ldots, k$. Prove that

$$\int_{\partial D} g \frac{f'}{f} dz = 2\pi i \left(\sum_{\ell=1}^{k} K_\ell g(b_\ell) - \sum_{j=1}^{n} N_j g(a_j) \right).$$

16.8. Use Exercise 16.7 to compute the integral

$$\int_{|z|=1} e^z \frac{4z^3 + 1}{z(2z^3 - 1)} dz.$$

16.9. Using Exercise 16.7, prove that

$$\frac{1}{2\pi i} \int_{|z|=4} \frac{9z^4 - 30z^3 + 12z^2 + 10z + 11}{z^3 - 6z^2 + 11z - 6} dz$$

is a positive integer. Can you compute it?

16.10. For each of the following paths, find $m \in \mathbb{Z}$ such that the path is homotopic in $\mathbb{C} \setminus \{0\}$ to $e^{2\pi i m t}$:

$$\text{(i) } (2 + \sin 2\pi t) e^{2\pi i t},$$

$$\text{(ii) } -(3 + \cos 4\pi t) e^{4\pi i t},$$

$$\text{(iii) } -\left(t^2 - t - \frac{3}{4} \right) e^{6\pi i \sin \frac{\pi t}{2}},$$

$$\text{(iv) } i \left(\left| t - \frac{1}{2} \right| + 1 \right) e^{2\pi i (4t^3 - 4t^2 + t)}.$$

Can you write an explicit formula for a CP-homotopy between each of the paths and the corresponding path $e^{2\pi i m t}$ in $\mathbb{C} \setminus \{0\}$?

Lecture 17

Index of a Path. The Argument Principle (Continued). Rouché's Theorem. Theorem 1.1 Revisited. Proof of Theorem 3.2. The Maximum Modulus Principle. Proof of Theorem 3.3

We continue to progress towards the standard version of the Argument Principle.

By Example 6.1, Theorem 9.1, and Proposition 16.2, for a closed path γ in $\mathbb{C} \setminus \{0\}$ the expression

$$\frac{1}{2\pi i} \int_\gamma \frac{1}{z} dz \qquad (17.1)$$

is an integer, namely, the integer m supplied by Proposition 16.2.

Definition 17.1. Let γ be a closed path γ in $\mathbb{C} \setminus \{0\}$. Then integer (17.1) is called *the index of γ [about 0]* and is denoted by $\mathrm{ind}_0 \gamma$.

Intuitively, $\mathrm{ind}_0 \gamma$ is "the number of times that γ goes around the origin", with the anti-clockwise direction corresponding to positive integers and the clockwise direction to negative ones.

As we will see shortly, the integral in the left-hand side of formula (16.8) can be expressed via the indices of certain paths.

Proposition 17.1. *Suppose that $D \subset \mathbb{C}$ is a bounded Jordan domain, with $\partial D = \sqcup_{j=1}^M \Gamma^j$, where Γ^j is the image of a piecewise C^1-smooth Jordan path for every j. Let γ^j be any parametrisation of Γ^j chosen in accordance with the standard orientation on Γ^j for every j. Suppose further that f is a function holomorphic on a neighbourhood of ∂D and having no zeroes in ∂D. Then*

$$\int_{\partial D} \frac{f'}{f} dz = 2\pi i \sum_{j=1}^M \mathrm{ind}_0(f \circ \gamma^j).$$

Proof. It suffices to prove

$$\int_{\Gamma^j} \frac{f'}{f} dz = 2\pi i \, \mathrm{ind}_0(f \circ \gamma^j), \quad j = 1, \ldots M. \qquad (17.2)$$

We assume that γ^j is a C^1-smooth path (hence a C^1-path) and leave the general case to the reader. Since $f \circ \gamma^j$ is a C^1-path as well, one has

$$\int_{\Gamma^j} \frac{f'}{f} dz = \int_{\gamma^j} \frac{f'}{f} dz = \int_0^1 \frac{f'(\gamma^j(t))}{f(\gamma^j(t))} \gamma^{j\prime}(t) dt =$$

$$\int_0^1 \frac{1}{(f \circ \gamma^j)(t)} (f \circ \gamma^j)'(t) dt = \int_{f \circ \gamma^j} \frac{1}{z} dz = 2\pi i \, \mathrm{ind}_0(f \circ \gamma^j)$$

as required in (17.2). □

Remark 17.1. Under the assumptions of Proposition 17.1, formula (17.2) shows that the number $\mathrm{ind}_0(f \circ \gamma^j)$ is independent of the choice of γ^j for every j, and we define

$$\mathrm{ind}_0(f \circ \Gamma^j) := \mathrm{ind}_0(f \circ \gamma^j), \quad j = 1, \ldots, M.$$

Further, define

$$\mathrm{ind}_0(f \circ \partial D) := \sum_{j=1}^M \mathrm{ind}_0(f \circ \Gamma^j).$$

Definition 17.2. Under the assumptions of Proposition 17.1, set

$$\Delta_{\partial D} f := 2\pi \, \mathrm{ind}_0(f \circ \partial D).$$

The number $\Delta_{\partial D} f$ is called *the change of the argument of f along ∂D.*

We are now ready to state the standard variant of the Argument Principle, which immediately follows from Theorem 16.1 and Proposition 17.1.

Theorem 17.1. (The Argument Principle, Final Version) *Let $D \subset \mathbb{C}$ be a bounded Jordan domain, $a_1, \ldots, a_n$ points in D, and $f \in H(\overline{D} \setminus \{a_1, \ldots, a_n\})$. Assume that f has a pole of order N_j at a_j, $j = 1, \ldots, n$. Suppose further that f does not vanish at any point of ∂D and let $b_1, \ldots, b_k$ be the zeroes of f in D. Define K_ℓ to be the order of b_ℓ, $\ell = 1, \ldots, k$. Then*

$$\Delta_{\partial D} f = 2\pi(K - N), \tag{17.3}$$

where $K := \sum_{\ell=1}^k K_\ell$ and $N := \sum_{j=1}^n N_j$.

Thus, the change of the argument of f along ∂D is fully determined by its zeroes and poles. We will now look at some examples.

Example 17.1. Set $D := \Delta$ and $f(z) := z^2$. Clearly, here $K = 2$, $N = 0$, so the right-hand side of formula (17.3) is equal to 4π. On the other hand, parametrising ∂D by the path $\gamma(t) := e^{2\pi i t}$, we see that $(f \circ \gamma)(t) = e^{4\pi i t}$. Therefore Example 6.1 yields $\mathrm{ind}_0(f \circ \gamma) = 2$, so the left-hand side of formula (17.3) is indeed equal to 4π.

Example 17.2. Let again $D := \Delta$ and

$$f(z) := 2z^2 - \frac{1}{z}.$$

This function has simple zeroes at the roots of $1/2$ of order 3 and a simple pole at 0, so $K = 3$ and $N = 1$, and the right-hand side of (17.3) is again equal to

4π. On the other hand, parametrising ∂D by the path $\gamma(t) := e^{2\pi it}$, we compute $(f \circ \gamma)(t) = 2e^{4\pi it} - e^{-2\pi it}$ and

$$\Delta_{\partial D} f = 2\pi \frac{1}{2\pi i} \int_{f \circ \gamma} \frac{1}{z} dz = -i \int_0^1 \frac{1}{2e^{4\pi it} - e^{-2\pi it}} (8\pi i e^{4\pi it} + 2\pi i e^{-2\pi it}) dt =$$

$$2\pi \int_0^1 \frac{(4e^{4\pi it} - 2e^{-2\pi it}) + 3e^{-2\pi it}}{2e^{4\pi it} - e^{-2\pi it}} dt = 4\pi + 6\pi \int_0^1 \frac{e^{-2\pi it}}{2e^{4\pi it} - e^{-2\pi it}} dt =$$

$$4\pi - 3i \int_{|z|=1} \frac{1}{z(2z^3 - 1)} dz = 4\pi$$

as expected, where the integral in the last line is equal to zero (explain!).

Let $D \subset \mathbb{C}$ be a bounded domain and $f \in H(\overline{D})$ with $f(z) \neq 0$ for all $z \in \partial D$. Then, by Theorem 12.2, the function f has only finitely many zeroes in D, and, analogously to the number K in formula (17.3), we denote by $K_{f,D}$ the number of zeroes of f in D where each zero is counted with its order; in other words, $K_{f,D}$ is the sum of the orders of all zeroes of f in D. We will now obtain a very useful consequence of Theorem 17.1.

Theorem 17.2. (Rouché's Theorem) *Let $D \subset \mathbb{C}$ be a bounded Jordan domain and $f \in H(\overline{D})$. Suppose that $g \in H(\overline{D})$ and $|g(z)| < |f(z)|$ for all $z \in \partial D$. Then $K_{f+g,D} = K_{f,D}$.*

In words, Theorem 17.2 states that the number of zeroes of a holomorphic function is stable under "sufficiently small" holomorphic perturbations.

Proof. Write $\partial D = \sqcup_{j=1}^M \Gamma^j$, where Γ^j is the image of a piecewise C^1-smooth Jordan path for every j, and let γ^j be any parametrisation of Γ^j chosen in accordance with the standard orientation on Γ^j, $j = 1, \ldots, M$. By Theorem 17.1, we have

$$K_{f+g,D} = \mathrm{ind}_0((f+g) \circ \partial D) = \sum_{j=1}^M \mathrm{ind}_0((f+g) \circ \gamma^j) = \frac{1}{2\pi i} \sum_{j=1}^M \int_{(f+g) \circ \gamma^j} \frac{1}{z} dz$$

and, analogously,

$$K_{f,D} = \frac{1}{2\pi i} \sum_{j=1}^M \int_{f \circ \gamma^j} \frac{1}{z} dz.$$

Thus, in order to prove the theorem, it suffices to show

$$\int_{(f+g) \circ \gamma^j} \frac{1}{z} dz = \int_{f \circ \gamma^j} \frac{1}{z} dz, \quad j = 1, \ldots, M. \tag{17.4}$$

Now, by the assumption $|g(z)| < |f(z)|$ for all $z \in \partial D$, the map

$$\Gamma(t,s) := f(\gamma^j(t)) + (1-s)g(\gamma^j(t))$$

is a CP-homotopy between $(f+g) \circ \gamma^j$ and $f \circ \gamma^j$ in $\mathbb{C} \setminus \{0\}$ (check!). Therefore, (17.4) holds by Theorem 9.1, and the proof is complete. $\quad\square$

As an application of Theorem 17.2, let us give a proof of the Fundamental Theorem of Algebra (Theorem 1.1) different from the one included in Lecture 1.

Proof (An alternative proof of Theorem 1.1). Without loss of generality we can assume that $a_K = 1$. As in the proof included in Lecture 1, write $P(z) = z^K + Q(z)$, where

$$Q(z) := a_{K-1} z^{K-1} + a_{K-2} z^{K-2} + \cdots + a_0,$$

and notice, as before, that $|Q(z)| < |z|^K$ if $|z| \geq R$ for all sufficiently large R. Applying Theorem 17.2 to $D := \Delta_R$, $f(z) := z^K$, $g(z) := Q(z)$ for such R, we see

$$K_{P, \Delta_R} = K_{z^K, \Delta_R} = K.$$

It is also clear that the complement $\mathbb{C} \setminus \Delta_R$ contains no zeroes of P. Thus, we have shown that P has exactly K zeroes in the complex plane, provided each zero is counted with its order. □

Theorem 17.2 is also the basis of the proof of the Open Mapping Theorem (Theorem 3.2), which is an important result formulated in Lecture 3 without proof. Essentially, it states that non-constant holomorphic functions are *open maps*, i.e., take open sets to open sets. We will now establish this theorem.

Proof (Theorem 3.2). We must prove that $f(D)$ is a domain. First of all, as D is path-connected, so is $f(D)$. Thus, we only need to see that $f(D) \subset \mathbb{C}$ is open.

Fix $w_0 \in f(D)$ and show that there exists a neighbourhood of w_0 in $\mathbb{C}$ lying in $f(D)$. Choose $z_0 \in f^{-1}(w_0)$ (so we have $f(z_0) = w_0$) and $r > 0$ such that:

(1) $\Delta(z_0, r) \subset D$;
(2) $f(z) \neq w_0 \ \forall z \in \overline{\Delta(z_0, r)} \setminus \{z_0\}$,

where to satisfy condition (2) we recall that $f \not\equiv \text{const}$ and appeal to Theorem 12.2 (explain!).

Now, set

$$\rho := \min_{|z-z_0|=r} |f(z) - w_0|$$

and notice that $\rho > 0$ by condition (2). We claim that the disk $\Delta(w_0, \rho)$ is contained in $f(D)$. Indeed, fix $w \in \Delta(w_0, \rho)$ and write

$$f(z) - w = (f(z) - w_0) + (w_0 - w).$$

Clearly, we have

$$|f(z) - w_0| \geq \rho > |w_0 - w| \ \forall z \in \partial \Delta(z_0, r).$$

Therefore, by Theorem 17.2 applied to the disk $\Delta(z_0, r)$ and the functions $f(z) - w_0$ and $w_0 - w$, we obtain

$$K_{f(z)-w, \Delta(z_0,r)} = K_{f(z)-w_0, \Delta(z_0,r)} \geq 1.$$

Hence, the function f takes the value w at some point of the disk $\Delta(z_0, r)$, and so $w \in f(D)$. Thus, we have shown that $\Delta(w_0, \rho) \subset f(D)$ as required. $\square$

Remark 17.2. Notice that Theorem 3.2 does not hold for $\mathbb{R}$-differentiable mappings from $\mathbb{R}^2$ to $\mathbb{R}^2$. For instance, the image of the map

$$F : \mathbb{R}^2 \to \mathbb{R}^2, \quad (x,y) \mapsto (x,x)$$

is the line $\{z \in \mathbb{C} : x = y\}$, which is not an open subset of $\mathbb{C}$.

Theorem 3.2 yields the following important result:

Theorem 17.3. (The Maximum Modulus Principle) *Let $D \subset \mathbb{C}$ be a domain, $f \in H(D)$, and suppose that there exist $z_0 \in D$ and $r > 0$ such that $\Delta(z_0, r) \subset D$ and $|f(z_0)| \geq |f(z)|$ for all $z \in \Delta(z_0, r)$. Then $f \equiv const$.*

In words, the modulus of a holomorphic function cannot attain a local maximum unless the function is constant.

Proof. We are given that $f(z_0)$ is a point having the greatest distance from the origin among all points $f(z)$ for $z \in \Delta(z_0, r)$. At the same time, assuming that $f \not\equiv const$ on $\Delta(z_0, r)$, we see by Theorem 3.2 that $f(\Delta(z_0, r))$ is an open subset of $\mathbb{C}$. In particular, $f(z_0)$ lies in the set $f(\Delta(z_0, r))$ together with a disk, say $\Delta(f(z_0), \rho)$. But $\Delta(f(z_0), \rho)$ contains a point, say $f(z_1)$, for which the distance from the origin is greater than that for $f(z_0)$ (see Fig. 17.1). This contradiction shows that $f \equiv const$ on $\Delta(z_0, r)$, and it then follows by Theorem 12.2 that $f \equiv const$ on D. $\square$

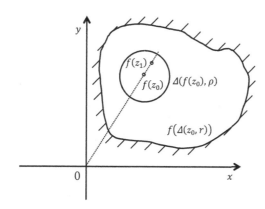

Fig. 17.1

Theorem 17.3 has a very useful consequence.

Corollary 17.1. *Let $D \subset \mathbb{C}$ be a bounded domain and $f \in H(D) \cap C(\overline{D})$. Then*

$$\max_{z \in \overline{D}} |f(z)| = \max_{z \in \partial D} |f(z)|. \tag{17.5}$$

In other words, the maximum of the modulus of a function holomorphic on a bounded domain and continuous on its closure is always attained on the boundary.

Proof. Let $z_0 \in \overline{D}$ be a point such that $|f(z_0)| \geq |f(z)|$ for all $z \in \overline{D}$. If $z_0 \in \partial D$, there is nothing to prove, so we assume that $z_0 \in D$. In this case, by Theorem 17.3 we have $f \equiv$ const on D. Hence $f \equiv$ const on $\overline{D}$, and thus (17.5) trivially holds. $\square$

Next, we will see that the general idea of the proof of Theorem 3.2 given above can be also used to establish Theorem 3.3 stated in Lecture 3.

Proof (Theorem 3.3). Suppose that $f'(z_0) = 0$ for some $z_0 \in D$ and let $w_0 := f(z_0)$. Choose $r > 0$ such that:

(1) $\Delta(z_0, r) \subset D$;
(2) $f'(z) \neq 0 \; \forall z \in \Delta(z_0, 0, r)$,

where to satisfy condition (2) we recall that $f \not\equiv$ const and use Theorem 12.2.
   As before, set

$$\rho := \min_{|z - z_0| = r} |f(z) - w_0|$$

and notice that $\rho > 0$. Fix $w \in \Delta(w_0, 0, \rho)$. Then

$$|f(z) - w_0| \geq \rho > |w_0 - w| \; \forall z \in \partial \Delta(z_0, r).$$

Since

$$f(z) - w = (f(z) - w_0) + (w_0 - w),$$

by Theorem 17.2 applied to the disk $\Delta(z_0, r)$ and the functions $f(z) - w_0$ and $w_0 - w$, we obtain

$$K_{f(z) - w, \Delta(z_0, r)} = K_{f(z) - w_0, \Delta(z_0, r)}.$$

As $f'(z_0) = 0$, we have $K_{f(z) - w_0, \Delta(z_0, r)} \geq 2$, hence $K_{f(z) - w, \Delta(z_0, r)} \geq 2$.
   Observe now that by condition (3) above, every zero of the function $f(z) - w$ in $\Delta(z_0, r)$ is simple, and therefore f takes the value w at least at two distinct point in $\Delta(z_0, 0, r)$. This contradicts the assumption that f is 1-to-1, and the proof is complete. $\square$

Exercises

17.1. For the path γ from Exercise 8.4, compute $\mathrm{ind}_0 \gamma$.

17.2. Let $D \subset \mathbb{C}$ be a bounded Jordan domain, $a_1, \ldots, a_n$ points in D, and $f \in H(\overline{D} \setminus \{a_1, \ldots, a_n\})$. Assume that f has a pole of order N_j at a_j, $j = 1, \ldots, n$.

Define $M := \max_{z \in \partial D} |f(z)|$ and fix $A \in \mathbb{C}$ with $|A| > M$. Let $b_1, \ldots, b_k$ be the zeroes of the function $f + A$ in D. Set K_ℓ to be the order of b_ℓ, $\ell = 1, \ldots, k$. Prove that

$$\sum_{\ell=1}^{k} K_\ell = \sum_{j=1}^{n} N_j.$$

17.3. Prove that the entire function $\sin z - z$ has infinitely many zeroes in $\mathbb{C}$. (Hint: find a sequence of rectangular non-intersecting domains $D_n \subset H$, $n \in \mathbb{N}$, with one of the sides of ∂D_n being the interval $[2\pi n, 2\pi(n+1)]$ and the length of the vertical sides varying depending on n, such that the change of the argument of $\sin z - z$ along ∂D_n is defined and positive for all n.)

17.4. Find the number of zeroes of the following polynomials in z in the respective domains (every zero must be counted with its order):

(i) $z^5 - 2z^3 + 5z - 1$ in the disk Δ,

(ii) $z^4 - 5z + 1$ in the disks Δ and Δ_2,

(iii) $z^3 - 5z + 1$ in the disks Δ, Δ_2 and Δ_3,

(iv) $z^4 - 12z + 2$ in the annulus $\Delta(0, 2, 3)$.

17.5. Let $P(z) = z^K + a_{K-1}z^{K-1} + \cdots + a_1 z + a_0$ be a polynomial in z of positive degree with leading coefficient equal to 1. Assume that $|a_j| \leq 1$, $j = 0, \ldots, K - 1$. Prove that all roots of P lie in the disk Δ_2.

17.6. Let $a > 1$. Prove that the equation $z e^{a-z} = 1$ has precisely one solution in the disk Δ.

17.7. Prove that for any $a \in \mathbb{C}$ and any integer $K \geq 2$ the polynomial $az^K + z + 1$ in z has at least one zero in the closed disk $\overline{\Delta}_2$. (Hint: use Vieta's formulas.)

17.8. Let $D \subset \mathbb{C}$ be a bounded Jordan domain and $f, g \in H(\overline{D})$. Assume that f has no zeroes in ∂D and one of the following holds:

(i) $\operatorname{Re} \dfrac{g(z)}{f(z)} > -1 \; \forall z \in \partial D$,

(ii) $\dfrac{g(z)}{f(z)} \in \mathbb{C} \setminus (-\infty, -1] \; \forall z \in \partial D.$

Prove that $K_{f+g,D} = K_{f,D}$.

17.9. Is Theorem 17.2 valid for unbounded Jordan domains? Prove your conclusion.

17.10. Find all entire functions $f = u + iv$ satisfying $u^3 - v^3 \equiv 1$.

17.11. Let $D \subset \mathbb{C}$ be a domain and $f \in H(D)$. Assume that $f = u + iv$ satisfies one of the following conditions: (i) $v \equiv$ const, (ii) $v = u^2$, (iii) $|f| \equiv$ const. Show that $f \equiv$ const.

17.12. Let $f \in H(\Delta(0,1,2)) \cap C(\overline{\Delta(0,1,2)})$. Write $f = u + iv$ and suppose that $u(z) = 0$ if $|z| = 1$ and $v(z) = 0$ if $|z| = 2$. Prove that $f \equiv 0$. (Hint: observe that the boundary of $f(\Delta(0,1,2))$ is contained in the image of the boundary of $\Delta(0,1,2)$ under f.)

17.13. Let $D \subset \mathbb{C}$ be a bounded domain and $\{f_n\}$ a sequence of functions with $f_n \in H(D) \cap C(\overline{D})$ for all n. Assume that $\{f_n\}$ converges uniformly on ∂D. Prove that $\{f_n\}$ converges to a function $g \in H(D) \cap C(\overline{D})$ uniformly on $\overline{D}$.

17.14. Let $f \in H(\Delta)$ with $|f(z)| < 1$ for all $z \in \Delta$. Assume that $f(\pm z_0) = 0$ for some non-zero $z_0 \in \Delta$. Prove that $|f(0)| \leq |z_0|^2$. (Hint: "divide" f by the two zeroes using suitable Möbius transformations of Δ.)

17.15. Let $D \subset \mathbb{C}$ be a bounded domain and $f \in H(D) \cap C(\overline{D})$. Suppose that $|f| \equiv$ const on ∂D and that f has no zeroes in D. Prove that $f \equiv$ const.

17.16. Let f be an entire function that maps the unit circle S^1 into itself. Prove that $f(z) = cz^K$ for some $c \in \mathbb{C}$ and some non-negative integer K. (Hint: reduce to Exercise 17.15 by "getting rid" of the zeroes of f in Δ.)

17.17. Assume that for a polynomial $P(z)$ in z and for some $M > 0$ we have

$$\max_{|z| \leq 1} |P(z)| \leq M.$$

Prove that for any $K \geq \deg P$ the following holds:

$$|P(z)| \leq M|z|^K \; \forall z \in \mathbb{C} \setminus \overline{\Delta}.$$

(Hint: consider the function $P(z)/z^K$ and transform it so that Corollary 17.1 becomes applicable to it.)

17.18. Prove that for a polynomial $P(z)$ in z the function

$$Q(r) := \frac{\max_{|z|=r} |P(z)|}{r^K}$$

is decreasing for every $K \geq \deg P$, and in fact strictly decreasing unless $P(z) = cz^K$ for some $c \in \mathbb{C}$. (Hint: the idea is similar to that for doing Exercise 17.17.)

17.19. Let $D \subset \mathbb{C}$ be a domain, $f, g \in H(D)$, and suppose that there exist $z_0 \in D$ and $r > 0$ such that $\Delta(z_0, r) \subset D$ and $|f(z_0)| + |g(z_0)| \geq |f(z)| + |g(z)|$ for all $z \in \Delta(z_0, r)$. Prove that $f \equiv$ const, $g \equiv$ const. (Hint: you may wish to apply Theorem 17.3 to the function $e^{e^{i\alpha}} f + e^{i\beta} g$ for suitable $\alpha, \beta \in \mathbb{R}$.)

17.20. Let $a \in \mathbb{C}$ be an essential singularity of a function f. Prove that

$$\lim_{\varepsilon \to 0} \varepsilon^n \max_{|z-a|=\varepsilon} |f(z)| = \infty \quad \forall n \in \mathbb{N}.$$

(Hint: use Theorem 14.2 and Corollary 17.1.)

Lecture 18

Schwarz's Lemma. Conformal Maps of the Unit Disk and the Upper Half-Plane. (Pre)-Compact Subsets of a Metric Space. Continuous Linear Functionals on $H(D)$. Arzelà-Ascoli's Theorem. Montel's Theorem. Hurwitz's Theorem

We will now derive another very useful corollary from Theorem 17.3.

Lemma 18.1. (Schwarz's Lemma) *Let $f \in H(\Delta)$ and $|f(z)| \leq 1$ for all $z \in \Delta$. Assume further that $f(0) = 0$. Then $|f(z)| \leq |z|$ for all $z \in \Delta$. Moreover, if for some non-zero $z_0 \in \Delta$ one has $|f(z_0)| = |z_0|$, then there exists $\alpha \in \mathbb{R}$ such that $f(z) = e^{i\alpha}z$.*

In particular, the above lemma states that under the map f the distance between a point in Δ and the origin is not increased. This fact and its generalisations have fundamental importance for complex analysis and geometry.

Proof. Consider

$$g(z) := \frac{f(z)}{z}, \ z \in \Delta \setminus \{0\}.$$

As $f(0) = 0$, the function g has a removable singularity at the origin, so by setting $g(0) := f'(0)$ we arrive at a function holomorphic on Δ (explain!).

Fix $z_0 \in \Delta$ and choose r satisfying $|z_0| < r < 1$. Then, by Corollary 17.1 applied to the function g and domain Δ_r, we have

$$|g(z_0)| \leq \max_{|z|=r} |g(z)| = \max_{|z|=r} \left| \frac{f(z)}{z} \right| \leq \frac{1}{r}.$$

By letting $r \to 1$, we obtain $|g(z_0)| \leq 1$, so $|f(z_0)| \leq |z_0|$ if $z_0 \neq 0$. Also note that this inequality trivially holds for $z_0 = 0$ by the assumption $f(0) = 0$, thus the first statement of the lemma is established.

On the other hand, if $|f(z_0)| = |z_0|$ for some $z_0 \in \Delta \setminus \{0\}$, we have $|g(z_0)| = 1$, and Theorem 17.3 implies $g \equiv e^{i\alpha}$, where $\alpha \in \mathbb{R}$. Hence, $f(z) = e^{i\alpha}z$, as required in the second statement of the lemma. $\square$

Next, recall that in Corollary 5.5 we found all Möbius transformations of Δ. Using Lemma 18.1 we will now show that any conformal transformation of Δ is a Möbius transformation and thus is given by formula (5.2).

Corollary 18.1. *Every conformal transformation f of Δ has the form (5.2), namely*

$$f(z) = e^{i\alpha} \frac{z-a}{1-\bar{a}z}$$

for some $a \in \Delta$ and $\alpha \in \mathbb{R}$.

Proof. By Theorem 4.1 one has $f \in H(\Delta)$. Set $b := f^{-1}(0)$ and consider the map

$$\lambda(z) := \frac{z+b}{1+\bar{b}z},$$

which is of the form (5.2) and therefore is a Möbius transformation of Δ. Then the composition $g := f \circ \lambda$ satisfies the assumptions of Lemma 18.1, so we have $|g(z)| \le |z|$ for all $z \in \Delta$.

Since g is a conformal transformation of Δ, by Theorem 4.2 one has $g^{-1} \in H(\Delta)$, and, applying Lemma 18.1 to g^{-1} one observes that $|g(z)| = |z|$ for all $z \in \Delta$. Utilising Lemma 18.1 once again, we see that $g(z) = e^{i\alpha}z$ for some $\alpha \in \mathbb{R}$. It then follows that $f(z) = e^{i\alpha}\lambda^{-1}(z)$, which completes the proof. □

Similarly, for conformal transformations of the other standard domain, the upper half-plane H, we have:

Corollary 18.2. *Any conformal transformation f of H has the form (5.3), namely*

$$f(z) = \frac{\mathbf{a}z+\mathbf{b}}{\mathbf{c}z+\mathbf{d}}$$

for some $\mathbf{a},\mathbf{b},\mathbf{c},\mathbf{d} \in \mathbb{R}$ with $\mathbf{ad}-\mathbf{bc} > 0$.

Proof. Homework. (Hint: refer to the proof of Proposition 5.6.) □

By Corollary 5.3 we now deduce:

Corollary 18.3. *The group of conformal transformations of each of Δ and H with respect to composition is isomorphic to* $\mathrm{PSL}_2(\mathbb{R}) = \mathrm{SL}_2(\mathbb{R}) \Big/ \left\{ \pm \begin{pmatrix} 1 & 0 \\ 0 & 1 \end{pmatrix} \right\}.$

The remainder of the course is aimed at establishing the Riemann Mapping Theorem (Theorem 8.3). Before proving it, we need to make certain preparations and obtain a number of independently interesting results. The first group of results is related to convergence of sequences of holomorphic functions.

Fix a domain $D \subset \mathbb{C}$. Recall from Lecture 11 that the function d defined in (11.3) turns the complex vector space $H(D)$ into a metric space, with the convergence coinciding with uniform convergence inside D. From now on, we will adapt the metric space terminology and refer to convergence inside D as *convergence in $H(D)$*.

Definition 18.1. Let X be a metric space. A subset $\mathscr{F} \subset X$ is called *compact* if any sequence in $\mathscr{F}$ has a subsequence converging to an element of $\mathscr{F}$.

In fact, the above definition refers to *sequential compactness* rather than compactness, but in metric spaces these two concepts coincide.

Definition 18.2. Let X be a metric space. A subset $\mathscr{F} \subset X$ is called *pre-compact* if any sequence in $\mathscr{F}$ has a convergent subsequence (with the limit not necessarily lying in $\mathscr{F}$).

One immediately observes that $\mathscr{F}$ is pre-compact if and only if its closure $\overline{\mathscr{F}}$ in the metric space X is compact (explain!).

Analogously to the fact that the supremum of a continuous real-valued function on a compact subset $\mathbf{K} \subset \mathbb{R}^n$ is attained at some point of $\mathbf{K}$, we have:

Proposition 18.1. *Let X be a metric space, $\mathscr{F} \subset X$ a compact subset and $\Phi : X \to \mathbb{C}$ a continuous map. Then there exists $y \in \mathscr{F}$ such that*

$$|\Phi(y)| = \sup_{x \in \mathscr{F}} |\Phi(x)|.$$

Remark 18.1. In the above proposition, the continuity of Φ is understood in the usual way as the continuity of a map between metric spaces. This is equivalent to the *sequential continuity* of Φ, i.e., to the condition that for any $x \in X$ and any sequence $\{x_n\}$ in X converging to x one has $\lim_{n \to \infty} \Phi(x_n) = \Phi(x)$.

Proof (Proposition 18.1). Let $A := \sup_{x \in \mathscr{F}} |\Phi(x)| \in \mathbb{R} \cup \{\infty\}$. Then there exists a sequence $\{x_n\} \subset \mathscr{F}$ such that

$$\lim_{n \to \infty} |\Phi(x_n)| = A.$$

Since $\mathscr{F}$ is compact, there exists a subsequence $\{x_{n_k}\}$ of $\{x_n\}$ converging to a point $y \in \mathscr{F}$. Then, by the continuity of Φ we have $\lim_{k \to \infty} |\Phi(x_{n_k})| = |\Phi(y)|$. Therefore, $A < \infty$ and $|\Phi(y)| = A$ as required. $\square$

One is often interested in continuous maps from $H(D)$ to $\mathbb{C}$ of a special kind.

Definition 18.3. A map $L : H(D) \to \mathbb{C}$ is called *a continuous linear functional on $H(D)$* if L is continuous and in addition is *linear*, where linearity is understood as the condition $L(ab + bg) = aL(f) + bL(g)$ for all $f, g \in H(D)$ and $a, b \in \mathbb{C}$.

We will now give an example of a continuous linear functional on $H(D)$ that will be of importance for us later on.

Example 18.1. Fix $z_0 \in D$ and $k \in \{0\} \cup \mathbb{N}$. Set

$$L(f) := f^{(k)}(z_0), \ f \in H(D).$$

Clearly, L is linear, so in order to show that L is a continuous linear functional on $H(D)$ we need to prove that L is continuous.

Fix $f \in H(D)$, and let $\{f_n\}$ be a sequence converging to f in $H(D)$. By Theorem 10.4, for all sufficiently small $\rho > 0$ we have

$$L(f_n) = \frac{k!}{2\pi i} \int_{|z-z_0|=\rho} \frac{f_n(z)}{(z-z_0)^{k+1}} dz,$$

and Lemma 11.1 implies

$$\lim_{n\to\infty} L(f_n) = \frac{k!}{2\pi i} \int_{|z-z_0|=\rho} \frac{f(z)}{(z-z_0)^{k+1}} dz = L(f),$$

which proves that L is indeed continuous (see also Theorem 11.7).

We will now attempt to come up with conditions that guarantee the pre-compactness of a subset of $H(D)$. First, we consider this problem for spaces of continuous maps on compact subsets in Euclidean space.

For every $\ell \in \mathbb{N}$, let $||\cdot||_\ell$ be the Euclidean norm on $\mathbb{R}^\ell$. If $\mathbf{K}$ is any a compact subset of $\mathbb{R}^n$, denote by $C(\mathbf{K}, \mathbb{R}^m)$ the real vector space of continuous maps from $\mathbf{K}$ to $\mathbb{R}^m$. Introduce a norm on $C(\mathbf{K}, \mathbb{R}^m)$ as follows:

$$||f||_{\mathbf{K}} := \max_{x \in \mathbf{K}} ||f(x)||_m, \ f \in C(\mathbf{K}, \mathbb{R}^m).$$

This norm induces a distance function on $C(\mathbf{K}, \mathbb{R}^m)$ in the usual way:

$$d_{\mathbf{K}}(f,g) := ||f-g||_{\mathbf{K}},$$

thus turning $C(\mathbf{K}, \mathbb{R}^m)$ into a metric space. Clearly, the convergence with respect to $d_{\mathbf{K}}$ is uniform convergence on $\mathbf{K}$.

Definition 18.4.
(1) A subset $\mathscr{F} \subset C(\mathbf{K}, \mathbb{R}^m)$ is said to be *equicontinuous* if for every $\varepsilon > 0$ there exists $\delta > 0$ such that $||f(x) - f(y)||_m < \varepsilon$ for all $f \in \mathscr{F}$ whenever for $x, y \in \mathbf{K}$ we have $||x - y||_n < \delta$.
(2) A subset $\mathscr{F} \subset C(\mathbf{K}, \mathbb{R}^m)$ is said to be *pointwise bounded* if for every $x \in \mathbf{K}$ there exists $M > 0$ such that $||f(x)||_m < M$ for all $f \in \mathscr{F}$.

We shall now state without proof a fundamental criterion for the pre-compactness of a subset of $C(\mathbf{K}, \mathbb{R}^m)$.

Theorem 18.1. (Arzelà-Ascoli's Theorem) *A subset $\mathscr{F} \subset C(\mathbf{K}, \mathbb{R}^m)$ is pre-compact if and only if $\mathscr{F}$ is both equicontinuous and pointwise bounded.*

We will use Theorem 18.1 to establish a result that will be required for our proof of Theorem 8.3.

Theorem 18.2. (Montel's Theorem) *Let $\mathscr{F} \subset H(D)$. Assume that for every compact subset $\mathbf{K} \subset D$ there exists $M > 0$ such that $|f(z)| < M$ for all $z \in \mathbf{K}$ and $f \in \mathscr{F}$. Then $\mathscr{F}$ is pre-compact.*

In words, a subset of $H(D)$ with locally uniformly bounded modulus is pre-compact.

Proof. For every compact subset $\mathbf{K} \subset D$, define $\mathscr{F}|_{\mathbf{K}}$ to be the subset of $C(\mathbf{K}, \mathbb{R}^2)$ that consists of the restrictions of the elements of $\mathscr{F}$ to $\mathbf{K}$. The main ingredient of the proof of the theorem is the following fact:

Lemma 18.2. $\mathscr{F}_{\mathbf{K}} \subset C(\mathbf{K}, \mathbb{R}^2)$ *is equicontinuous for every compact subset* $\mathbf{K} \subset D$.

We postpone the proof of Lemma 18.2 for a moment and establish the theorem assuming that the lemma holds.

Consider the exhaustion of D by the compact subsets $\mathbf{K}_n$ introduced in (11.1). Lemma 18.2 yields that $\mathscr{F}_{\mathbf{K}_n} \subset C(\mathbf{K}_n, \mathbb{R}^2)$ is equicontinuous for every $n \in \mathbb{N}$. Hence, as $\mathscr{F}_{\mathbf{K}_n}$ is pointwise bounded by assumption, Theorem 18.1 implies that this subset is pre-compact in $C(\mathbf{K}_n, \mathbb{R}^2)$.

Fix a sequence $\{f_m\}$ in the set $\mathscr{F}$ and look at the sequence of restrictions $\left\{ f_m|_{\mathbf{K}_1} \right\} \subset C(\mathbf{K}_1, \mathbb{R}^2)$. Since $\mathscr{F}_{\mathbf{K}_1}$ is pre-compact in $C(\mathbf{K}_1, \mathbb{R}^2)$, one can find a sub-sequence $\left\{ f_{m_{\ell_1}} \Big|_{\mathbf{K}_1} \right\}$ of $\left\{ f_m|_{\mathbf{K}_1} \right\}$ convergent in $C(\mathbf{K}_1, \mathbb{R}^2)$. Next, consider the sub-sequence $\left\{ f_{m_{\ell_1}} \right\}$ of $\{f_m\}$ and the sequence of restrictions $\left\{ f_{m_{\ell_1}} \Big|_{\mathbf{K}_2} \right\} \subset C(\mathbf{K}_2, \mathbb{R}^2)$. Again, by the pre-compactness of $\mathscr{F}_{\mathbf{K}_2}$ in $C(\mathbf{K}_2, \mathbb{R}^2)$, one can choose a subsequence $\left\{ f_{m_{\ell_1 \ell_2}} \Big|_{\mathbf{K}_2} \right\}$ of $\left\{ f_{m_{\ell_1}} \Big|_{\mathbf{K}_2} \right\}$ convergent in $C(\mathbf{K}_2, \mathbb{R}^2)$. Continuing in this way, for every $n \in \mathbb{N}$ we build a subsequence

$$\left\{ f_{m_{\ell_1}}_{\cdots_{\ell_n}} \right\} \subset \{f_m\}$$

that converges uniformly on the compact subset $\mathbf{K}_n$. Now, we choose *the diagonal subsequence*

$$\left\{ f_{m_1}, f_{m_{\ell_{1_2}}}, \ldots, f_{m_{\ell_{1_{\ell_2}}}}_{\cdots_n}, \ldots \right\},$$

which is constructed by taking the first element of the first subsequence, the second element of the second one, etc. The diagonal subsequence converges uniformly on $\mathbf{K}_n$ for every n and therefore converges in $H(D)$ (see formula (11.3)). This shows that $\mathscr{F}$ is pre-compact in $H(D)$ as required. $\quad\square$

We will now obtain Lemma 18.2.

Proof (Lemma 18.2). Fix a compact subset $\mathbf{K} \subset D$ and choose a positive number $r < \mathrm{dist}(\mathbf{K}, \partial D)$. Define

$$\mathbf{K}^r := \{z \in \mathbb{C} : \mathrm{dist}(z, \mathbf{K}) \le r\}.$$

Clearly, $\mathbf{K}^r$ is a compact subset of D containing $\mathbf{K}$.

Let $M > 0$ be such that $|f(z)| < M$ for all $z \in \mathbf{K}^r$ and $f \in \mathscr{F}$. Fix $z_0 \in \mathbf{K}$, $f \in \mathscr{F}$ and set

$$g(z) := \frac{f(z_0 + rz) - f(z_0)}{2M}, \quad z \in \Delta.$$

Observe that $g \in H(\Delta)$, $|g(z)| \leq 1$ for all $z \in \Delta$, and $g(0) = 0$. Then, by Lemma 18.1, we deduce $|g(z)| \leq |z|$ for all $z \in \Delta$, i.e.,

$$|f(z_0 + rz) - f(z_0)| \leq 2M|z| \ \forall z \in \Delta.$$

Hence

$$|f(z) - f(z_0)| \leq \frac{2M}{r}|z - z_0| \ \forall z \in \Delta(z_0, r) \cap \mathbf{K}.$$

As z_0 is an arbitrary point of $\mathbf{K}$ and f is any element of $\mathscr{F}$, the equicontinuity of $\mathscr{F}_\mathbf{K}$ follows. □

The next fact that we will need for our proof of Theorem 8.3 is:

Theorem 18.3. (Hurwitz's Theorem) *Let $\{f_n\}$ be a sequence in $H(D)$ that converges in $H(D)$ to a function f. Assume that $f \not\equiv const$ and $f(z_0) = 0$ for some $z_0 \in D$. Then for every disk $\Delta(z_0, r) \subset D$ there exists $N \in \mathbb{N}$ such that the function f_n has a zero in $\Delta(z_0, r)$ for all $n \geq N$.*

In words, if the limit of a sequence of holomorphic function is non-constant and has a zero, then all elements of the sequence (with the exception of finitely many) vanish arbitrarily close to the zero of the limit.

Proof. By Theorem 12.2, there exists a positive number r_0 for which $\Delta(z_0, r_0) \subset D$ and f has no zeroes in $\Delta(z_0, r_0) \setminus \{z_0\}$. Clearly, it suffices to establish the theorem only for $r < r_0$.

Choose r as above and let

$$\varepsilon := \min_{z \in \partial \Delta(z_0, r)} |f(z)|.$$

Observe that $\varepsilon > 0$ by our choice of r. Since $\{f_n\}$ converges to f uniformly on $\partial \Delta(z_0, r)$, there exists $N \in \mathbb{N}$ such that

$$|f_n(z) - f(z)| < \varepsilon \ \forall z \in \partial \Delta(z_0, r), \ \forall n \geq N.$$

Then we have

$$|f_n(z) - f(z)| < |f(z)| \ \forall z \in \partial \Delta(z_0, r), \ \forall n \geq N.$$

Therefore, writing $f_n = f + (f_n - f)$, we see by Theorem 17.2 that for $n \geq N$

$$K_{f_n, \Delta(z_0, r)} = K_{f, \Delta(z_0, r)} \geq 1,$$

and the proof is complete. □

Corollary 18.4. *Let $\{f_n\}$ be a sequence in $H(D)$ that converges in $H(D)$ to a function f. Assume that $f \not\equiv const$ and that f_n is 1-to-1 for every n. Then f is 1-to-1.*

That is, the limit of a sequence of 1-to-1 holomorphic functions is either 1-to-1 or constant.

Proof. Suppose that f is not 1-to-1, so there exist $z_1, z_2 \in D$, with $z_1 \neq z_2$ and $f(z_1) = f(z_2)$. Consider

$$g_n(z) := f_n(z) - f_n(z_2).$$

The sequence $\{g_n\}$ converges in $H(D)$ to the function

$$g(z) := f(z) - f(z_2).$$

Clearly, g is non-constant and has a zero at z_1.

Fix a disk $\Delta(z_1, r) \subset D$ not containing the point z_2. Then by Theorem 18.3 there exists $N \in \mathbb{N}$ such that g_n has a zero, say w_n, in $\Delta(z_1, r)$ for all $n \geq N$. Hence, we have

$$f_n(w_n) - f_n(z_2) = 0 \ \forall n \geq N,$$

which is impossible as every function f_n is 1-to-1. $\quad\square$

Exercises

18.1. Let $f \in H(\Delta)$ and $|f(z)| \leq 1$ for all $z \in \Delta$. Assume that for some $K \in \mathbb{N}$ we have $f(0) = f'(0) = \ldots = f^{(K-1)}(0) = 0$. Prove that $|f(z)| \leq |z|^K$ for all $z \in \Delta$. Moreover, if for some non-zero $z_0 \in \Delta$ one has $|f(z_0)| = |z_0|^K$, then there exists $\alpha \in \mathbb{R}$ such that $f(z) = e^{i\alpha} z^K$.

18.2. Let $f \in H(\Delta)$ and assume that for some $M > 0$ we have $|f(z)| \leq M$ for all $z \in \Delta$. Prove that

$$|f'(z)| \leq \frac{M - \dfrac{1}{M}|f(z)|^2}{1 - |z|^2} \ \forall z \in \Delta.$$

(Hint: reduce to Lemma 18.1 by composing with suitable Möbius transformations of Δ.)

18.3. Let $f \in H(\Delta)$, $|f(z)| \leq 1$ for all $z \in \Delta$, and suppose that $f(0) = 0$, $f'(0) = 1$. Prove that $f(z) = z$.

18.4. Let $f \in H(\Delta)$, $|f(z)| \leq 1$ for all $z \in \Delta$, and suppose that f fixes two points of Δ. Prove that $f(z) = z$.

18.5. Let $f \in H(\Delta)$, $|\operatorname{Re} f(z)| \leq 1$ for all $z \in \Delta$, and suppose that $f(0) = 0$. Prove that $|f'(0)| \leq 4/\pi$. (Hint: reduce to Lemma 18.1 by composing with a suitable Möbius transformation.)

18.6. Let $f \in H(\Delta)$, $|f(z)| < 1$ for all $z \in \Delta$, and suppose that $f(0) = 0$. Define f_n to be the composition of n copies of f and suppose that $\{f_n\}$ converges to a function g on Δ. Prove that either $g \equiv 0$ or $g(z) = z$. (Hint: use Lemma 18.1.)

18.7. Let $D \subset \mathbb{C}$ be a domain and $\{f_n\}$ a sequence in $H(D)$ such that for some $M > 0$ one has $|f_n(z)| \leq M$ for all $z \in D$ and all n. Suppose that $E \subset D$ has a limit point in D and that $\{f_n\}$ converges on E. Prove that $\{f_n\}$ converges in $H(D)$. (Hint: application of Lemma 18.1 as in the proof of Lemma 18.2 may be useful.)

18.8. Let $\mathscr{F} \subset H(\Delta)$ be the subset of all conformal transformations of Δ. Show that in every sequence $\{f_n\}$ in $\mathscr{F}$ one can find a subsequence that converges in $H(\Delta)$ to a function $f \in H(\Delta)$ such that either (i) $f \in \mathscr{F}$, or (ii) $f \equiv c$, where $|c| = 1$.

18.9. Let

$$\mathscr{F} := \{f \in H(S_{-1,1}) : f(0) = 0 \text{ and } |f(z)| < 1 \ \forall z \in S_{-1,1}\},$$

where $S_{-1,1}$ is the strip as introduced in Lecture 4. Consider the continuous linear functional on $H(S_{-1,1})$ defined by

$$L(f) := f(1), \quad f \in H(S_{-1,1}).$$

Find $\sup_{f \in \mathscr{F}} |L(f)|$. Is this supremum attained at some point in $\mathscr{F}$? If so, can you find such a point? Is the set $\mathscr{F}$ compact in $H(S_{-1,1})$?

18.10. Let $D \subset \mathbb{C}$ be a domain, $\mathbf{K} \subset D$ an infinite compact subset and $\mathscr{F} \subset H(D)$. Let, further, $\mathscr{F}|_{\mathbf{K}}$ be the subset of $C(\mathbf{K}, \mathbb{R}^2)$ that consists of the restrictions of the elements of $\mathscr{F}$ to $\mathbf{K}$. Prove that the restriction map

$$R_{\mathbf{K}} : \mathscr{F} \to \mathscr{F}_{\mathbf{K}}, \quad f \mapsto f|_{\mathbf{K}}$$

is injective.

18.11. For a subset $S \subset \mathbb{C}$, define its *diameter* as

$$\operatorname{diam} S = \sup_{z_1, z_2 \in S} |z_1 - z_2|.$$

Fix $M > 0$ and let

$$\mathscr{F} := \{f \in H(\Delta) : f(0) = 0 \text{ and } \operatorname{diam} f(\Delta) \leq M\}.$$

Prove that $\mathscr{F}$ is compact in $H(\Delta)$.

18.12. Let $D \subset \mathbb{C}$ be a domain and $\mathscr{F} \subset H(D)$. Assume that for every $f \in \mathscr{F}$ and every $z \in D$ we have $f(z) \in \mathbb{C} \setminus \{z \in \mathbb{C} : x = y, x \geq 0\}$. Prove that for every sequence $\{f_n\}$ in $\mathscr{F}$ one can find a subsequence $\{f_{n_k}\}$ of $\{f_n\}$ such that either (i) $\{f_{n_k}\}$ converges in $H(D)$, or (ii) $\lim_{k \to \infty} f_{n_k}(z) = \infty$ for all $z \in D$. (Hint: use a suitable Möbius transformation and apply Theorem 18.2.)

18.13. Let $D \subset \mathbb{C}$ be a domain and $\{f_n\}$ a sequence in $H(D)$ such that f_n has no zeroes in D for all n and there exists $M > 0$ with $|f_n(z)| \leq M$ for all $z \in D$ and all n. Assume that for some $z_0 \in D$ the sequence $\{f_n(z_0)\}$ converges to 0. Prove that there exists a subsequence of $\{f_n\}$ that converges to 0 in $H(D)$.

18.14. Let $D \subset \mathbb{C}$ be a bounded domain and $\{f_n\}$ a sequence of conformal transformations of D. Assume that for some $z_0 \in D$ the sequence $\{f_n(z_0)\}$ converges to $a \in \partial D$. Prove that there is a subsequence of $\{f_n\}$ that converges to a in $H(D)$.

Lecture 19
Analytic Continuation

The second group of results that we need to obtain Theorem 8.3 concerns analytic continuation. In fact, as explained in Remark 20.1, we could avoid using analytic continuation for the purposes of proving Theorem 8.3. However, this is an important concept that is widely used in complex analysis in general, and we feel that it should be part of our presentation.

Definition 19.1. An *analytic element* is a pair (f, D), where D is an open disk of positive (possibly infinite) radius and f a function holomorphic on D. The centre and the radius of D are called *the centre of* (f, D) and *the radius of* (f, D), respectively. An analytic element is called *canonical* if D is the disk of convergence of the power series expansion of f centred at the centre of D. Two analytic elements (f, D) and $(\widetilde{f}, \widetilde{D})$ are said to be *equal* (we write $(f, D) = (\widetilde{f}, \widetilde{D})$) if $D = \widetilde{D}$ and $f(z) = \widetilde{f}(z)$ everywhere.

Definition 19.2. An analytic element $(\widetilde{f}, \widetilde{D})$ is said to be *an immediate analytic continuation* (IAC) *of an analytic element* (f, D) if $\widetilde{D} \cap D \neq \emptyset$ and $\widetilde{f}(z) = f(z)$ for all $z \in \widetilde{D} \cap D$.
Clearly, $(\widetilde{f}, \widetilde{D})$ is an IAC of (f, D) if and only if (f, D) is an IAC of $(\widetilde{f}, \widetilde{D})$, i.e., the condition of being an IAC is a symmetric binary relation. Therefore, in this situation we often say that (f, D) and $(\widetilde{f}, \widetilde{D})$ are *IACs of each other*.

We will now establish some properties of IACs.
Proposition 19.1.

(1) *Let two analytic elements* (f, D) *and* $(\widetilde{f}, \widetilde{D})$ *be IACs of each other and let the centre of* $(\widetilde{f}, \widetilde{D})$, *say* b, *lie in* D. *Then* $\widetilde{f}$ *is the restriction to* $\widetilde{D}$ *of the sum of the power series expansion of* f *with centre* b.

(2) *Suppose that for three analytic elements* (f_1, D_1), (f_2, D_2), (f_3, D_3) *we have* $D_1 \cap D_2 \cap D_3 \neq \emptyset$. *Assume that* (f_1, D_1) *and* (f_2, D_2) *are IACs of each other and that* (f_2, D_2) *and* (f_3, D_3) *are IACs of each other. Then* (f_1, D_1) *and* (f_3, D_3) *are IACs of each other as well.*

Proof. To obtain Part (1), consider the power series expansion of f with centre b:

$$\sum_{n=0}^{\infty} d_n(z-b)^n. \tag{19.1}$$

Clearly, this expansion converges to f on the disk $\Delta(b,r)$ with $r := \text{dist}(b, \partial D)$. Since $(\widetilde{f}, \widetilde{D})$ is an IAC of (f,D), one has

$$\widetilde{f}(z) = \sum_{n=0}^{\infty} d_n(z-b)^n \; \forall z \in \Delta(b,r) \cap \widetilde{D}.$$

Therefore, (19.1) is the power series expansion of $\widetilde{f}$ with centre b on the disk $\widetilde{D}$.

To obtain Part (2), notice that $f_1(z) = f_2(z) = f_3(z)$ on $D_1 \cap D_2 \cap D_3$, so $f_1(z) = f_3(z)$ on $D_1 \cap D_3$ by Theorem 12.2. $\square$

Remark 19.1. One way to arrive at the power series expansion of f with centre b in Part (1) of Proposition 19.1 is *to re-expand about b* the power series expansion of f centred at the centre of D, say a. The re-expansion process is as follows. Firstly, write the power series expansion of f with centre a

$$f(z) = \sum_{n=0}^{\infty} c_n(z-a)^n. \tag{19.2}$$

Secondly, write $z - a = (b-a) + (z-b)$ and, consequently,

$$(z-a)^n = \sum_{j=0}^{n} \binom{n}{j} (b-a)^j (z-b)^{n-j} \tag{19.3}$$

for all $n \in \mathbb{N}$. Substituting (19.3) into (19.2) and using Theorem 11.4, we obtain a power series with centre b converging to f on $\Delta(b, \text{dist}(b, \partial D))$ (provide details!).

We will now introduce the main player of this lecture starting with the more intuitive "discrete" variant of analytic continuation.

Definition 19.3. An analytic element $(\widetilde{f}, \widetilde{D})$ is said to be *an analytic continuation (AC) of an analytic element* (f,D) if there exist finitely many analytic elements $(\mathbf{f}_1, \mathbf{D}_1), \ldots, (\mathbf{f}_n, \mathbf{D}_n)$, with $(\mathbf{f}_1, \mathbf{D}_1) = (f,D)$, $(\mathbf{f}_n, \mathbf{D}_n) = (\widetilde{f}, \widetilde{D})$, such that $(\mathbf{f}_j, \mathbf{D}_j)$ and $(\mathbf{f}_{j+1}, \mathbf{D}_{j+1})$ are IACs of each other for all $j = 1, \ldots, n-1$.

Clearly, $(\widetilde{f}, \widetilde{D})$ is an AC of (f,D) if and only if (f,D) is an AC of $(\widetilde{f}, \widetilde{D})$, i.e., the condition of being an AC is a symmetric binary relation. Therefore, in this case we often say that (f,D) *and* $(\widetilde{f}, \widetilde{D})$ *are ACs of each other*.

To illustrate the above definition, let us recall Example 7.1. It is easy to observe from the example that the analytic elements constructed there yield that $(\widetilde{\ln}_0 + 2\pi i, \Delta^0)$ is an AC of $(\widetilde{\ln}_0, \Delta^0)$, where $\widetilde{\ln}_0$ was defined in Example 7.1. Note that all the analytic elements in Example 7.1 are canonical (explain!).

Often, it will be more convenient to utilise a "continuous" version of analytic continuation, which we will now introduce for canonical analytic elements.

Definition 19.4. Let γ be a path and (f,D) a canonical analytic element with centre $\gamma(0)$. *An analytic continuation (AC) of (f,D) along γ* is a family (f_t, D_t) of canonical analytic elements, with $t \in [0,1]$, such that

(1) $(f_0, D_0) = (f,D)$;
(2) the centre of (f_t, D_t) is $\gamma(t)$ for all $t \in [0,1]$;
(3) for every $t \in [0,1]$ there exists a connected neighbourhood U_t of t in $[0,1]$ (i.e., a subinterval of $[0,1]$ containing t) such that: (i) $\gamma(U_t) \subset D_t$, and (ii) (f_t, D_t) and (f_τ, D_τ) are IACs of each other for all $\tau \in U_t$.

If an AC of (f,D) along γ exists, we say that (f,D) *admits an AC along γ*.

Remark 19.2. In the above definition, the conditions that U_t is connected and $\gamma(U_t) \subset D_t$ can be omitted as any neighbourhood of a point $t \in [0,1]$ contains a neighbourhood of t with these properties.

We will now give two examples of an AC along a path. The first one can be thought of as "a continuous analogue" of the construction presented in Example 7.1.

Example 19.1. Let $D := \Delta(1,1)$ and

$$f(z) := \widetilde{\mathrm{ln}}_0 z, \; z \in D.$$

The function f is the inverse to e^z on a certain subdomain of the strip $S_{-\pi,\pi} = \{z \in \mathbb{C} : -\pi < \mathrm{Im}\, z < \pi\}$ (cf. the proof of Proposition 4.2) and therefore is holomorphic on D. Consider the canonical analytic element (f,D) and let $\gamma(t) := e^{2\pi i t}$. To construct an AC of (f,D) along γ, set

$$D_t := \Delta(e^{2\pi i t}, 1)$$

and

$$f_t(z) := \ln|z| + i\widetilde{\mathrm{arg}}_t z, \; z \in D_t,$$

where $\widetilde{\mathrm{arg}}_t$ is the analogue of arg measured from $2\pi t - \pi$ to $2\pi t + \pi$ rather than from 0 to 2π (so for the function $\widetilde{\mathrm{arg}}$ introduced in Example 7.1 we have $\widetilde{\mathrm{arg}} = \widetilde{\mathrm{arg}}_0$). Arguing as in the proof of Proposition 4.2, one can show that the function f_t is the inverse to e^z on a subdomain of the strip

$$S_{2\pi t - \pi, 2\pi t + \pi} = \{z \in \mathbb{C} : 2\pi t - \pi < \mathrm{Im}\, z < 2\pi t + \pi\}$$

(explain!) and therefore is holomorphic on D_t. Then the family (f_t, D_t) of canonical analytic elements defines an AC of (f,D) along γ (prove!).

A completely analogous procedure yields an AC of (f,D) along any path in $\mathbb{C} \setminus \{0\}$ with initial point 1 (provide details!).

The second example is similar to the first one but involves roots in place of logarithms.

Example 19.2. Let $D := \Delta(1,1)$ and

$$f(z) := \sqrt{|z|}e^{i\frac{\widetilde{\arg z}}{2}}, \ z \in D.$$

As in the proof of Proposition 4.3, one can show that the function f is the inverse to z^2 on a subdomain of the right half-plane $\{z \in \mathbb{C} : \operatorname{Re} z > 0\}$ (explain!) and therefore is holomorphic on D. Consider the canonical analytic element (f,D) and let, again, $\gamma(t) := e^{2\pi i t}$. To build an AC of (f,D) along γ, set

$$D_t := \Delta(e^{2\pi i t},1)$$

and

$$f_t(z) := \sqrt{|z|}e^{i\frac{\widetilde{\arg}_t z}{2}}, \ z \in D_t,$$

where $\widetilde{\arg}_t$ is defined in Example 19.1. Arguing as in the proof of Proposition 4.3, one observes that the function f_t is the inverse to z^2 on a subdomain of the half-plane $A_{\pi t - \pi/2, \pi t + \pi/2}$ (explain!) and therefore is holomorphic on D_t. Then the family (f_t, D_t) of canonical analytic elements yields an AC of (f,D) along γ (prove!).

Again, an analogous construction leads to an AC of (f,D) along any path in $\mathbb{C} \setminus \{0\}$ with initial point 1 (provide details!).

We will now show that an AC along a path, if exists, is unique.

Proposition 19.2. *Let γ be a path and (f,D) a canonical analytic element with centre $\gamma(0)$. Let, further, (f_t, D_t) and $(\widetilde{f}_t, \widetilde{D}_t)$ be two ACs of (f,D) along γ. Then $(f_t, D_t) = (\widetilde{f}_t, \widetilde{D}_t)$ for all $t \in [0,1]$.*

Proof. Set

$$E := \left\{ t \in [0,1] : (f_t, D_t) = (\widetilde{f}_t, \widetilde{D}_t) \right\}.$$

As E contains 0, we have $E \neq \emptyset$. We shall show that E is both open and closed in $[0,1]$.

Choose U_t and $\widetilde{U}_t$ to be neighbourhoods of $t \in [0,1]$ corresponding to (f_t, D_t) and $(\widetilde{f}_t, \widetilde{D}_t)$, respectively, as in Definition 19.4. To see that E is open, fix $\tau_0 \in E$ and prove that the open subset $U := U_{\tau_0} \cap \widetilde{U}_{\tau_0}$ of $[0,1]$ lies in E. Fix $t \in U$. Then (f_{τ_0}, D_{τ_0}) and (f_t, D_t) are IACs of each other and (f_{τ_0}, D_{τ_0}) and $(\widetilde{f}_t, \widetilde{D}_t)$ are IACs of each other. Furthermore, as the centre of each of (f_t, D_t) and $(\widetilde{f}_t, \widetilde{D}_t)$ is $\gamma(t) \in D_{\tau_0}$, by Part (1) of Proposition 19.1 both (f_t, D_t) and $(\widetilde{f}_t, \widetilde{D}_t)$ are obtained from the power series expansion of f_{τ_0} with centre $\gamma(t)$. As the analytic elements (f_t, D_t) and $(\widetilde{f}_t, \widetilde{D}_t)$ are canonical, it follows that they are equal. Alternatively, we could refer to Part (2) of Proposition 19.1 to conclude that (f_t, D_t) and $(\widetilde{f}_t, \widetilde{D}_t)$ are IACs of each other and therefore are equal. Thus, $t \in U$, which shows that $U \subset E$. Hence, E is indeed open.

Next, to see that E is closed, consider a sequence $\{\sigma_n\}$ in E converging to $\tau_0 \in [0,1]$. As before, set $U := U_{\tau_0} \cap \widetilde{U}_{\tau_0}$. Then there exists $N \in \mathbb{N}$ such that $\sigma_n \in U$ for all $n \geq N$. Clearly, for all $n \geq N$ we have that $(f_{\sigma_n}, D_{\sigma_n})$ and (f_{τ_0}, D_{τ_0}) are IACs of each other as well as that $(f_{\sigma_n}, D_{\sigma_n})$ and $(\widetilde{f}_{\tau_0}, \widetilde{D}_{\tau_0})$ are IACs of each other. Furthermore, for all $n \geq N$ the intersection $D_{\sigma_n} \cap D_{\tau_0} \cap \widetilde{D}_{\tau_0}$ is non-empty as it contains the point $\gamma(\sigma_n)$. Therefore, by Part (2) of Proposition 19.1 we see that (f_{τ_0}, D_{τ_0}) and $(\widetilde{f}_{\tau_0}, \widetilde{D}_{\tau_0})$ are IACs of each other. As these analytic elements are canonical and have a common centre, they in fact coincide, so $\tau_0 \in E$. This proves that E is closed.

Since E is a non-empty open and closed subset of $[0,1]$, it follows that $E = [0,1]$ as required. $\square$

From now on, we will speak of *the* AC of a canonical analytic element (f, D) along a path γ (provided (f, D) admits *an* AC along γ).

Definition 19.5. Let γ be a path and (f, D) a canonical analytic element with centre $\gamma(0)$. Suppose that (f, D) admits an AC along γ, and let (f_t, D_t) be the AC of (f, D) along γ. Then the canonical analytic element (f_1, D_1) is called *the result of the AC of (f, D) along γ*.

Next, we shall record an important fact.

Proposition 19.3. *Assume that a canonical analytic element admits an AC along a path γ and let (f_t, D_t) be the AC of the analytic element along γ with R_t being the radius of (f_t, D_t). Then we have either* (i) $R_t < \infty$ *for all* $t \in [0,1]$ *and* $t \mapsto R_t$ *is a continuous function on* $[0,1]$, *or* (ii) $R_t = \infty$ *for all* $t \in [0,1]$.

Proof. Suppose first that $R_t < \infty$ for all $t \in [0,1]$. Fix $\tau_0 \in [0,1]$ and consider its neighbourhood U_{τ_0} as in Definition 19.4. We claim that for any $t \in U_{\tau_0}$ the intersection $\partial D_{\tau_0} \cap \partial D_t$ is non-empty (see Fig. 19.1). Indeed, otherwise the closure of one of the disks D_{τ_0}, D_t would lie in the other disk, which is impossible by Part (1) of Proposition 19.1.

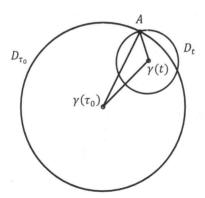

Fig. 19.1

Fix $t \in U_{\tau_0}$ and consider the triangle with vertices A, $\gamma(\tau_0)$, $\gamma(t)$ as shown in Fig. 19.1 (note that the triangle degenerates into a segment if $\partial D_{\tau_0} \cap \partial D_t$ consists of a single point). From this triangle we see

$$|R_{\tau_0} - R_t| \leq |\gamma(\tau_0) - \gamma(t)|.$$

The above inequality yields that the function $t \mapsto R_t$ is continuous at τ_0 as stated in (i).

Suppose next that the radius of one of the analytic elements in the family (f_t, D_t) is infinite and let

$$E := \{t \in [0,1] : R_t = \infty\}.$$

The set E is non-empty, and we will now show that E is both open and closed in $[0,1]$. Fix $\tau_0 \in E$ and look at its neighbourhood U_{τ_0} as in Definition 19.4. Then, by Part (1) of Proposition 19.1, for every $t \in U_{\tau_0}$ the analytic element (f_t, D_t) is obtained from the power series expansion of f_{τ_0} with centre $\gamma(t)$. It follows that $R_t = \infty$, thus $t \in E$, hence $U_{\tau_0} \subset E$, which proves that E is open.

Next, to see that E is closed, let $\{\sigma_n\}$ be a sequence in E converging to a point $\tau_0 \in [0,1]$ and consider a neighbourhood U_{τ_0} of τ_0 as in Definition 19.4. Then there exists $N \in \mathbb{N}$ such that $\sigma_n \in U_{\tau_0}$ for all $n \geq N$. Therefore, $(f_{\sigma_n}, D_{\sigma_n})$ and (f_{τ_0}, D_{τ_0}) are IACs of each other for all $n \geq N$. As $R_{\sigma_n} = \infty$, the point $\gamma(\tau_0)$ lies in D_{σ_n}. Hence, by Part (1) of Proposition 19.1, the analytic element (f_{τ_0}, D_{τ_0}) is obtained from the power series expansion of f_{σ_n} with centre $\gamma(\tau_0)$, which implies $R_{\tau_0} = \infty$, thus $\tau_0 \in E$. This proves that E is closed.

Since E is a non-empty open and closed subset of $[0,1]$, we see that $E = [0,1]$ as stated in (ii). $\square$

Exercises

19.1. Is the following analytic element canonical:

$$\left(\sum_{n=2}^{\infty} \frac{z^n}{n \ln n}, \Delta_{1/2} \right) ?$$

Prove your conclusion.

19.2. Are the analytic elements

$$\left(\sum_{n=0}^{\infty} \frac{z^n}{n!}, \Delta_2 \right) \qquad \text{and} \qquad \left(\sum_{n=0}^{\infty} \frac{e(z-1)^n}{n!}, \Delta(1,2) \right)$$

IACs of each other? Prove your conclusion.

19.3. Re-expand about $1/2$ and $i/2$ the geometric series $\sum_{n=0}^{\infty} z^n$. What are the radii of the canonical analytic elements arising from the re-expanded series? Prove your conclusion.

19.4. Show that the analytic elements (f, D) and (f_1, D_1) from Example 19.2 are ACs of each other.

19.5. Adapt Example 19.2 to roots of order n. Namely, let $D := \Delta(1,1)$ and

$$f(z) := \sqrt[n]{|z|}e^{i\frac{\overline{\arg}z}{n}}, \ z \in D.$$

Construct an AC of (f,D) along $\gamma(t) := e^{2\pi i t}$.

19.6. Does the canonical analytic element

$$\left(\sum_{n=1}^{\infty} z^{n!}, \Delta\right)$$

admit an AC along the path $\gamma(t) := 2it$? Prove your conclusion. (Hint: recall Exercise 11.10.)

19.7. Does the canonical analytic element

$$\left(\sum_{n=1}^{\infty} z^{2^n}, \Delta\right)$$

admit an AC along the path $\gamma(t) := 3te^{\pi i t}$? Prove your conclusion.

19.8. Find the result of the AC of the analytic element (f,D) from Example 19.2 along the path γ from Exercise 8.4.

19.9. Assume that a canonical analytic element (f,D) admits an AC along a path γ and that $(\widetilde{f},\widetilde{D})$ is the result of the AC of (f,D) along γ. Prove that $(\widetilde{f},\widetilde{D})$ admits an AC along the path γ_- and that (f,D) is the result of the AC of $(\widetilde{f},\widetilde{D})$ along γ_-.

19.10. Let (f,D) be a canonical analytic element that admits an AC along a path γ. Suppose that the result of the AC of (f,D) along γ is $(0,\mathbb{C})$. Prove that $f \equiv 0$.

19.11. Construct an AC of the analytic element (f,D) from Example 19.2 along the path

$$\gamma(t) := \left(1 + \frac{1}{2}\sin 2\pi t\right)e^{6\pi i t}$$

and find the corresponding function $t \mapsto R_t$.

19.12. Consider the analytic element

$$\left(\sum_{n=0}^{\infty} \frac{(z-3)^{3n}}{(n!)^2}, D\right),$$

where D is the disk of convergence of the power series $\displaystyle\sum_{n=0}^{\infty} \frac{(z-3)^{3n}}{(n!)^2}$. Does this analytic element admit an AC along the path $\gamma(t) := e^{t^2+2\pi i t^3}$? If so, find the corresponding function $t \mapsto R_t$. (Hint: you may find Stirling's approximation useful.)

Lecture 20
Analytic Continuation (Continued). The Monodromy Theorem

We will now show that the "discrete" and "continuous" variants of analytic continuation given in Definitions 19.3 and 19.4, respectively, are closely related. The fact stated below is not required for our proof of Theorem 8.3 later in the course, but we have chosen to include it in the lecture in order to further clarify the concept of analytic continuation.

Proposition 20.1.

(1) *Suppose that a canonical analytic element (f,D) admits an AC along a path γ and let $(\widetilde{f},\widetilde{D})$ be the result of the AC of (f,D) along γ. Then $(\widetilde{f},\widetilde{D})$ is an AC of (f,D).*

(2) *Conversely, if a canonical analytic element $(\widetilde{f},\widetilde{D})$ is an AC of a canonical analytic element (f,D), then there exists a path, say γ, such that (f,D) admits an AC along γ and $(\widetilde{f},\widetilde{D})$ is the result of the AC of (f,D) along γ.*

Proof. To establish Part (1), let (f_t, D_t) be the AC of (f,D) along γ, where $(f_1,D_1) = (\widetilde{f},\widetilde{D})$, and let R_t be the radius of (f_t,D_t). Proposition 19.3 implies that there exists $\varepsilon > 0$ with $R_t \geq \varepsilon$ for all $t \in [0,1]$. By the uniform continuity of γ, there is $\delta > 0$ such that $|\gamma(t) - \gamma(t')| < \varepsilon$ whenever we have $|t - t'| < \delta$ for $t,t' \in [0,1]$. Choose a partition $0 = t_0 < t_1 < \cdots < t_n = 1$ of the segment $[0,1]$ with $|t_j - t_{j+1}| < \delta$ for $j = 0,\ldots,n-1$. Part (1) of the proposition is then a consequence of the lemma stated below.

Lemma 20.1. *For $j = 0,\ldots,n-1$ the analytic elements (f_{t_j}, D_{t_j}) and $(f_{t_{j+1}}, D_{t_{j+1}})$ are IACs of each other.*

Proof. Fix $j \in \{0,\ldots,n-1\}$. As for any two points $t,t' \in [t_j, t_{j+1}]$ one has

$$|\gamma(t) - \gamma(t')| < \varepsilon \leq \min\{R_t, R_{t'}\},$$

it follows that $\gamma([t_j,t_{j+1}]) \subset D_t$ for all $t \in [t_j, t_{j+1}]$. In particular, for $t,t',t'' \in [t_j, t_{j+1}]$ we have $D_t \cap D_{t'} \cap D_{t''} \neq \emptyset$.

175

Now, for every $t \in [t_j, t_{j+1}]$ consider the interval $U_t \cap [t_j, t_{j+1}]$, where U_t is a neighbourhood of t as in Definition 19.4. These sets form an open cover of $[t_j, t_{j+1}]$, and we choose a finite subcover, say $\mathcal{U}$, of this cover. Let I_1 be an interval in $\mathcal{U}$ containing the point t_j. Then, by Part (2) of Proposition 19.1, the analytic elements (f_{t_j}, D_{t_j}) and (f_t, D_t) are IACs of each other for every $t \in I_1$. If the point t_{j+1} lies in I_1, the lemma follows, so we assume that $I_1 = [t_j, \tau_0)$, with $\tau_0 \leq t_{j+1}$. In this case, find an interval $I_2 \in \mathcal{U}$ containing the point τ_0. Then (f_t, D_t) and $(f_{t'}, D_{t'})$ are IACs of each other for every $t \in I_2$ and every $t' \in I_1 \cap I_2$. Therefore, (f_{t_j}, D_{t_j}) and (f_t, D_t) are IACs of each other for every $t \in I_2$. Again, if t_{j+1} lies in I_2, the lemma follows, otherwise we choose an interval $I_3 \in \mathcal{U}$ containing the point $\sup I_2$. Proceeding in this way, we will eventually reach an interval $I \in \mathcal{U}$ that contains t_{j+1} such that (f_{t_j}, D_{t_j}) and (f_t, D_t) are IACs of each other for every $t \in I$. The proof of the lemma is now complete. $\square$

Next, to obtain Part (2), let $(\mathbf{f}_1, \mathbf{D}_1), \ldots, (\mathbf{f}_n, \mathbf{D}_n)$ be analytic elements with $(\mathbf{f}_1, \mathbf{D}_1) = (f, D)$, $(\mathbf{f}_n, \mathbf{D}_n) = (\widetilde{f}, \widetilde{D})$ and such that $(\mathbf{f}_j, \mathbf{D}_j)$ and $(\mathbf{f}_{j+1}, \mathbf{D}_{j+1})$ are IACs of each other for all $j = 1, \ldots, n-1$. Define γ as the collection of the $n-1$ segments joining the centres of $(\mathbf{f}_1, \mathbf{D}_1), \ldots, (\mathbf{f}_n, \mathbf{D}_n)$. More precisely, if a_j is the centre of $(\mathbf{f}_j, \mathbf{D}_j)$, we set

$$\gamma(t) := \left(1 - ((n-1)t - j)\right) a_{j+1} + \left(((n-1)t - j)\right) a_{j+2}$$
$$\text{if } \frac{j}{n-1} \leq t \leq \frac{j+1}{n-1}, \; j = 0, \ldots, n-2.$$

Next, for $j/(n-1) \leq t \leq (j+1)/(n-1)$ introduce (f_t, D_t) as follows. If $\gamma(t)$ lies in $\mathbf{D}_j$ (resp. in $\mathbf{D}_{j+1}$) then we take f_t to be the sum of the power series expansion of $\mathbf{f}_j$ (resp. of $\mathbf{f}_{j+1}$) centred at $\gamma(t)$ and D_t to be the disk of convergence of this expansion. Notice that for $\gamma(t) \in \mathbf{D}_j \cap \mathbf{D}_{j+1}$, each of the two expansions yields the same canonical analytic element (explain!). We obviously have $(f_0, D_0) = (f, D)$ and $(f_1, D_1) = (\widetilde{f}, \widetilde{D})$. Next, if $\gamma(t)$ lies in $\mathbf{D}_j$ (resp. in $\mathbf{D}_{j+1}$), we take U_t to be any connected neighbourhood of t in $[0, 1]$ such that $\gamma(U_t) \subset \mathbf{D}_j \cap D_t$ (resp. such that $\gamma(U_t) \subset \mathbf{D}_{j+1} \cap D_t$). Then, using Part (2) of Proposition 19.1, we see that (f_t, D_t) and (f_τ, D_τ) are IACs of each other for all $\tau \in U_t$ (provide details!).

Thus, we have shown that (f_t, D_t) is an AC of (f, D) along γ and that $(\widetilde{f}, \widetilde{D})$ is the result of the AC of (f, D) along γ as required. $\square$

We will now obtain our main technical fact on analytic continuation.

Theorem 20.1. *Let γ and $\tilde{\gamma}$ be two paths with common endpoints and (f, D) a canonical analytic element with centre $\gamma(0) = \tilde{\gamma}(0)$. Let, further, $\Gamma(t, s)$ be a PCE-homotopy between γ and $\tilde{\gamma}$ in $\mathbb{C}$ and assume that (f, D) admits an AC along every path $\gamma_s(t) := \Gamma(t, s)$, $s \in [0, 1]$. Then the results of the ACs of (f, D) along γ and $\tilde{\gamma}$ coincide.*

Proof. For every $s \in [0, 1]$, denote the AC of (f, D) along γ_s by $(f_{t,s}, D_{t,s})$, with $R_{t,s}$ being the radius of the canonical analytic element $(f_{t,s}, D_{t,s})$. Our goal is to show that $(f_{1,0}, D_{1,0}) = (f_{1,1}, D_{1,1})$. Notice that it suffices to prove that for every $s \in [0,1]$ there exists a neighbourhood V_s of s in $[0,1]$ such that for all $s' \in V_s$ one has $(f_{1,s'}, D_{1,s'}) = (f_{1,s}, D_{1,s})$. Indeed, in this case the map

$$s \mapsto (f_{1,s}, D_{1,s})$$

from $[0,1]$ to the set of all analytic elements with centre at the terminal point $\gamma(1) = \tilde{\gamma}(1)$ would be locally constant, and therefore constant, which implies $(f_{1,0}, D_{1,0}) = (f_{1,1}, D_{1,1})$ as desired.

We now fix $s_0 \in [0,1]$ and construct a required neighbourhood V_{s_0}. By Proposition 19.3, there exists $\varepsilon > 0$ such that

$$R_{t,s_0} \geq \varepsilon \ \forall t \in [0,1]. \tag{20.1}$$

We then set V_{s_0} to be any neighbourhood of s_0 in $[0,1]$ for which

$$|\gamma_{s_0}(t) - \gamma_s(t)| < \frac{\varepsilon}{4} \ \forall t \in [0,1] \tag{20.2}$$

whenever $s \in V_{s_0}$; the existence of such a neighbourhood is guaranteed by the uniform continuity of the function Γ. For every $s \in V_{s_0}$ and $t \in [0,1]$ we now introduce a canonical analytic element $(\widehat{f}_{t,s}, \widehat{D}_{t,s})$ with centre $\gamma_s(t)$ by setting $\widehat{f}_{t,s}$ to be the sum of the power series expansion of f_{t,s_0} centred at $\gamma_s(t)$ and $\widehat{D}_{t,s}$ to be the disk of convergence of this expansion. Clearly, we have

$$(\widehat{f}_{0,s}, \widehat{D}_{0,s}) = (f_{0,s_0}, D_{0,s_0}) = (f,D) \text{ and } (\widehat{f}_{1,s}, \widehat{D}_{1,s}) = (f_{1,s_0}, D_{1,s_0}) \text{ for all } s \in V_{s_0}.$$

Furthermore, by (20.1), (20.2) the radius $\widehat{R}_{t,s}$ of $(\widehat{f}_{t,s}, \widehat{D}_{t,s})$ satisfies

$$\widehat{R}_{t,s} \geq \frac{3\varepsilon}{4}. \tag{20.3}$$

Now, in order to see that for all $s \in V_{s_0}$ one has $(f_{1,s}, D_{1,s}) = (f_{1,s_0}, D_{1,s_0})$, we only need to prove that for every $s \in V_{s_0}$ the family $(\widehat{f}_{t,s}, \widehat{D}_{t,s})$ is an AC of (f,D) along γ_s. Indeed, by Proposition 19.2 we would then have

$$(f_{1,s}, D_{1,s}) = (\widehat{f}_{1,s}, \widehat{D}_{1,s}) = (f_{1,s_0}, D_{1,s_0}).$$

Fix $s \in V_{s_0}$. To show that the family $(\widehat{f}_{t,s}, \widehat{D}_{t,s})$ is an AC of (f,D) along γ_s, choose $\delta > 0$ such that

$$|\gamma_{s_0}(t) - \gamma_{s_0}(t')| < \frac{\varepsilon}{4} \tag{20.4}$$

for all $t,t' \in [0,1]$ satisfying $|t - t'| < \delta$; the existence of δ with this property follows from the uniform continuity of Γ. Let

$$\widehat{U}_{t,s} := U_{t,s_0} \cap (t - \delta, t + \delta) \ \forall t \in [0,1],$$

where U_{t,s_0} is a neighbourhood of t in $[0,1]$ corresponding to the family (f_{t,s_0}, D_{t,s_0}) as in Definition 19.4. We will now fix $t \in [0,1]$ and prove that $(\widehat{f}_{\tau,s}, \widehat{D}_{\tau,s})$ and $(\widehat{f}_{t,s}, \widehat{D}_{t,s})$ are IACs of each other for all $\tau \in \widehat{U}_{t,s}$.

First of all, notice that the intersection $D_{t,s_0} \cap D_{\tau,s_0} \cap \widehat{D}_{t,s}$ is non-empty as by estimates (20.1), (20.2), (20.3), (20.4) it contains the points $\gamma_{s_0}(t)$, $\gamma_{s_0}(\tau)$, $\gamma_s(t)$ (see Fig. 20.1). Furthermore, (f_{t,s_0}, D_{t,s_0}) and $(f_{\tau,s_0}, D_{\tau,s_0})$ are IACs of each other as $\tau \in U_{t,s_0}$. Also, (f_{t,s_0}, D_{t,s_0}) and $(\widehat{f}_{t,s}, \widehat{D}_{t,s})$ are IACs of each other by the construction of $(\widehat{f}_{t,s}, \widehat{D}_{t,s})$. Therefore, Part (2) of Proposition 19.1 implies that $(f_{\tau,s_0}, D_{\tau,s_0})$ and $(\widehat{f}_{t,s}, \widehat{D}_{t,s})$ are IACs of each other.

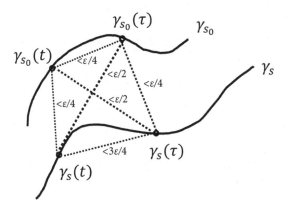

Fig. 20.1

Next, observe that by estimates (20.1), (20.2), (20.3), (20.4) the intersection $D_{\tau,s_0} \cap \widehat{D}_{t,s} \cap \widehat{D}_{\tau,s}$ contains the points $\gamma_{s_0}(\tau)$, $\gamma_s(t)$, $\gamma_s(\tau)$ and therefore is non-empty (see Fig. 20.1). Moreover, as we have just shown, $(f_{\tau,s_0}, D_{\tau,s_0})$ and $(\widehat{f}_{t,s}, \widehat{D}_{t,s})$ are IACs of each other, and $(f_{\tau,s_0}, D_{\tau,s_0})$ and $(\widehat{f}_{\tau,s}, \widehat{D}_{\tau,s})$ are IACs of each other by the definition of $(\widehat{f}_{t,s}, \widehat{D}_{t,s})$. Hence, Part (2) of Proposition 19.1 yields that $(\widehat{f}_{t,s}, \widehat{D}_{t,s})$ and $(\widehat{f}_{\tau,s}, \widehat{D}_{\tau,s})$ are IACs of each other.

Finally, as we have already noted, $\gamma_s(\tau) \in \widehat{D}_{t,s}$ for all $t \in [0,1]$ and $\tau \in \widehat{U}_{t,s}$. We have thus proved that the family $(\widehat{f}_{t,s}, \widehat{D}_{t,s})$ is an AC of (f,D) along γ_s as required. $\square$

Theorem 20.1 leads to one of the most important facts concerning analytic continuation.

Corollary 20.1. (The Monodromy Theorem) *Let $G \subset \mathbb{C}$ be a simply-connected domain and (f,D) a canonical analytic element with centre $a \in G$. Assume that (f,D) admits an AC along any path in G with initial point a. Then there exists a function $F \in H(G)$ such that $F(z) = f(z)$ for all $z \in \Delta(a, \mathrm{dist}(a, \partial G)) \cap D$. In other words, $f|_{\Delta(a,\mathrm{dist}(a,\partial G)) \cap D}$ can be holomorphically extended to G.*

Specifically, in the proof below we will show that if a canonical analytic element admits an analytic continuation along any path (with the initial point being the centre of the element) lying in a simply-connected domain, then the results of the analytic continuations along all the paths define a function holomorphic on the domain.

Proof. Fix $z_0 \in G$ and let γ be a path in G joining a and z_0 (so we have $\gamma(0) = a$ and $\gamma(1) = z_0$). Consider the AC of (f, D) along γ, say (f_t, D_t), and set

$$F(z_0) := f_1(z_0).$$

As G is simply-connected, Theorem 20.1 implies that this definition is independent of the choice of γ, and we obtain a well-defined function on G.

To see that $F \in H(G)$, we shall prove that F coincides with f_1 on the disk $V := \Delta(z_0, \mathrm{dist}(z_0, \partial G)) \cap D_1$. Indeed, fix $z \in V$ and consider the path $\tilde{\gamma}$ in G obtained by "augmenting" γ with the segment $[z_0, z]$. More precisely, set

$$\tilde{\gamma}(t) := \begin{cases} \gamma(2t) & \text{if } 0 \le t \le 1/2, \\ 2(z - z_0)t + 2z_0 - z & \text{if } 1/2 \le t \le 1. \end{cases}$$

We claim that the AC of (f, D) along $\tilde{\gamma}$ is obtained by joining the family (f_t, D_t) with the canonical analytic elements derived from the power series expansions of f_1 with centres at the points of $[z_0, z]$. Namely, we claim that the AC of (f, D) along $\tilde{\gamma}$ is given by

$$(\tilde{f}_t, \tilde{D}_t) := \begin{cases} (f_{2t}, D_{2t}) & \text{if } 0 \le t \le 1/2, \\ (\hat{f}_t, \hat{D}_t) & \text{if } 1/2 \le t \le 1, \end{cases}$$

where $\hat{f}_t$ is the sum of the power series expansion of the function f_1 with centre $2(z - z_0)t + 2z_0 - z$ and $\hat{D}_t$ is the disk of convergence of this expansion.

Indeed, choose a neighbourhood $\tilde{U}_t$ of $t \in [0, 1]$ as follows:

(1) for $t \in [0, 1/2)$ set $\tilde{U}_t := \dfrac{1}{2} U_{2t}$, where U_τ is a neighbourhood of $\tau \in [0, 1]$ corresponding to the family (f_τ, D_τ) as in Definition 19.4;

(2) set $\tilde{U}_{1/2} := \dfrac{1}{2} U_1 \cup (1/2, 1]$;

(3) for $t \in (1/2, 1]$ set $\tilde{U}_t$ to be any connected neighbourhood of t with $\tilde{\gamma}(\tilde{U}_t) \subset \tilde{D}_t$.

Then, it is not hard to see that $(\tilde{f}_t, \tilde{D}_t)$ and $(\tilde{f}_\tau, \tilde{D}_\tau)$ are IACs of each other for all $t \in [0, 1]$ and $\tau \in \tilde{U}_t$ (provide details!). Furthermore, it is easy to observe that $\tilde{\gamma}(\tilde{U}_t) \subset \tilde{D}_t$ for all $t \in [0, 1]$.

We have thus shown that $(\tilde{f}_t, \tilde{D}_t)$ is the AC of (f, D) along $\tilde{\gamma}$. Therefore, by the definition of F, one has

$$F(z) = \tilde{f}_1(z) = f_1(z).$$

Hence, F coincides with f_1 on V, which proves that $F \in H(G)$.

It remains to see that F is equal to f on the disk $\Delta(a, \mathrm{dist}(a, \partial G)) \cap D$. Indeed, analogously to the family $(\hat{f}_t, \hat{D}_t)$ for $1/2 \le t \le 1$ considered above, for

$z \in \Delta(a, \text{dist}(a, \partial G)) \cap D$ the AC of (f, D) along the segment $[a, z]$ is given by the power series expansions of f with centres at the points of $[a, z]$ (provide details!). Therefore, $F(z) = f(z)$ as required. $\square$

For our proof of Theorem 8.3 we will need the following result, which is based on Corollary 20.1:

Proposition 20.2. *Let $G \subset \mathbb{C}$ be a simply-connected domain and $f \in H(G)$ a function that does not vanish at any point of G. Then there exists $g \in H(G)$ such that $g^2 = f$.*

Proof. Fix $z_0 \in G$ and choose (F, D) to be a canonical analytic element with centre $f(z_0) \in \mathbb{C} \setminus \{0\}$ and $D \subset \mathbb{C} \setminus \{0\}$ satisfying $F^2(z) = z$ everywhere (explain why such an analytic element exists!). Consider the open subset

$$V := f^{-1}(D \cap f(G)) \subset G,$$

which contains z_0, and let $\widetilde{D}$ be any disk with centre z_0 lying in V. Clearly, we have $F \circ f|_{\widetilde{D}} \in H(\widetilde{D})$. Now, let h be the sum of the power series expansion of $F \circ f|_{\widetilde{D}}$ with centre z_0 and $\widehat{D}$ the disk of convergence of the expansion. Consider the canonical analytic element $(h, \widehat{D})$. In order to prove the theorem, we will show that $(h, \widehat{D})$ admits an AC along any path γ in G with $\gamma(0) = z_0$. Indeed, in this case by Corollary 20.1 we would obtain a function, say g, holomorphic on G and equal to $F \circ f$ on a $\widetilde{D}$. Then g^2 coincides with f on $\widetilde{D}$, which by Theorem 12.2 implies that $g^2 = f$ on G as required.

Choose any path γ in G with $\gamma(0) = z_0$ and set $\tilde{\gamma} := f \circ \gamma$. Certainly, $\tilde{\gamma}$ is a path in $\mathbb{C} \setminus \{0\}$ with $\tilde{\gamma}(0) = f(z_0)$. As in Example 19.2, the analytic element (F, D) admits an AC along any path in $\mathbb{C} \setminus \{0\}$ with initial point $f(z_0)$ and let (F_t, D_t) be the AC of (F, D) along $\tilde{\gamma}$. For every $t \in [0, 1]$ consider the open set

$$V_t := f^{-1}(D_t \cap f(G)) \subset G,$$

which contains the point $\gamma(t)$, and choose a disk $\widetilde{D}_t$ with centre $\gamma(t)$ lying in V_t. We have $F_t \circ f|_{\widetilde{D}_t} \in H(\widetilde{D}_t)$. Let $\widehat{D}_t$ be the disk of convergence of the sum, say h_t, of the power series expansion of $F_t \circ f|_{\widetilde{D}_t}$ with centre $\gamma(t)$ and look at the canonical analytic element $(h_t, \widehat{D}_t)$. We claim that the family $(h_t, \widehat{D}_t)$ is an AC of $(h, \widehat{D})$ along γ.

First of all, we clearly have $(h_0, \widehat{D}_0) = (h, \widehat{D})$. Next, for every $t \in [0, 1]$ choose $\delta > 0$ such that $\gamma(t') \in \widetilde{D}_t$ for all $t' \in [0, 1]$ satisfying $|t - t'| < \delta$. Set

$$\widehat{U}_t := U_t \cap (t - \delta, t + \delta),$$

where U_t is a neighbourhood of t in $[0, 1]$ corresponding to the family (F_t, D_t) as in Definition 19.4. Fix $\tau \in \widehat{U}_t$ and show that $(h_t, \widehat{D}_t)$ and $(h_\tau, \widehat{D}_\tau)$ are IACs of each other. By Theorem 12.2, it suffices to prove that h_τ coincides with h_t on the intersection $W := \widetilde{D}_\tau \cap \widetilde{D}_t$, which is non-empty as $\gamma(\tau) \in \widetilde{D}_t$.

On W we have $h_t = F_t \circ f$, $h_\tau = F_\tau \circ f$. But $f(W) \subset D_t \cap D_\tau$ and, since $\tau \in U_t$, we see that $F_\tau = F_t$ on $D_t \cap D_\tau$. It then follows that $h_\tau = h_t$ on W, hence $(h_t, \widehat{D}_t)$ and $(h_\tau, \widehat{D}_\tau)$ are IACs of each other, and therefore $(h_t, \widehat{D}_t)$ is an AC of $(h, \widehat{D})$ along γ as claimed. $\square$

Remark 20.1. We will now give an alternative proof of Proposition 20.2, which does not rely on the theory of analytic continuation. By Corollary 9.2, there exists a holomorphic primitive, say F, of the function $f'/f \in H(G)$. By adding a constant to F if necessary, we can choose F to satisfy

$$e^{F(z_0)} = f(z_0) \tag{20.5}$$

for a fixed point $z_0 \in G$. Define

$$g(z) := e^{\frac{1}{2}F(z)} \ \forall z \in G.$$

To check that $g^2 = f$, we observe that $g^2 = e^F$, and therefore

$$(g^2)'(z) = e^{F(z)}\frac{f'(z)}{f(z)} = g^2(z)\frac{f'(z)}{f(z)} \ \forall z \in G.$$

Hence

$$(g^2)'f - g^2 f' \equiv 0,$$

which implies

$$\frac{g^2}{f} \equiv \text{const},$$

and by condition (20.5) we see that $g^2 = f$ as required. $\square$

Exercises

20.1. Prove that the analytic elements

$$\left(\sum_{n=0}^{\infty} z^n, \Delta\right) \quad \text{and} \quad \left(\sum_{n=0}^{\infty} (-1)^{n+1}(z-2)^n, \Delta(2,1)\right)$$

are canonical and that they are ACs of each other. Furthermore, find a path γ with $\gamma(0) = 0$ such that the first analytic element admits an AC along γ and the result of the AC of this analytic element along γ is the second analytic element.

20.2. Do Exercise 19.8 using Theorem 20.1.

20.3. Let (f, D) a canonical analytic element with centre a that admits an AC along any path in $\mathbb{C}$ with initial point a. Assume that the result of the AC of (f, D) along any such path has modulus not exceeding 1 at every point. Prove that $f \equiv \text{const}$. (Hint: use Theorem 13.1.)

20.4. Explain why there does not exist $g \in H(\mathbb{C} \setminus \{0\})$ with $g^2(z) = z$.

20.5. Explain why there does not exist $g \in H(\mathbb{C} \setminus \{0\})$ with $e^{g(z)} = z$.

20.6. Adapt each of the two proofs of Proposition 20.2 to roots of order n for any $n \geq 2$. Namely, let $G \subset \mathbb{C}$ be a simply-connected domain and $f \in H(G)$ a function that does not vanish at any point of G; show, following each of the two proofs of Proposition 20.2, that there exists $g \in H(G)$ such that $g^n = f$.

20.7. Let $G \subset \mathbb{C}$ be a simply-connected domain and $f \in H(G)$ a function that does not vanish at any point of G. Prove that there exists $g \in H(G)$ such that $e^g = f$. Give two arguments corresponding to the two proofs of Proposition 20.2.

20.8. Show that the functions

$$f(z) := \sum_{n=1}^{\infty} nz^n \text{ and } g(z) := \sum_{n=1}^{\infty} \frac{1}{n} z^n, \ z \in \Delta,$$

holomorphically extend to the domain $\mathbb{C} \setminus [1, \infty)$.

20.9. Is the following statement correct: if $f \in H(\Delta)$ and there exists $M > 0$ such that $|f(z)| \leq M$ for all $z \in \Delta$, then f holomorphically extends to a neighbourhood of $\overline{\Delta}$? Prove your conclusion.

20.10. A set $E \subset \mathbb{C}$ is said to be of *Hausdorff length zero* if for every $\varepsilon > 0$ it can be covered by countably many open disks for which the sum of the radii does not exceed ε. Let $G \subset \mathbb{C}$ be a domain and $E \subset G$ a compact subset of Hausdorff length zero. Suppose that $f \in H(G \setminus E)$ and there exists $M > 0$ such that $|f(z)| \leq M$ for all $z \in G \setminus E$. Prove that f holomorphically extends to G. Can you obtain this result if E is assumed to be closed in G rather than compact? (Hint: cover the set E by disks with arbitrarily small sum of the radii and use Theorem 10.4 for the resulting domain.)

20.11. Let (f, D) a canonical analytic element with centre $a \neq 0$ that admits an AC along any path in $\mathbb{C} \setminus \{0\}$ with initial point a. Assume that the result of the AC of (f, D) along any such path is independent of the path and has modulus not exceeding 1 at every point. Prove that $f \equiv$ const. (Hint: use Exercise 20.10.)

Lecture 21
Proof of Theorem 8.3. Conformal Transformations of the Canonical Simply-Connected Domains

We are now ready to prove the Riemann Mapping Theorem.

Proof (Theorem 8.3). First of all, by applying a suitable Möbius transformation we can assume that one of the points in $\overline{\mathbb{C}} \setminus D$ is ∞, i.e., that D lies in $\mathbb{C}$. As D is simply-connected, the set $\mathbb{C} \setminus D$ cannot be a single point (prove!), so we choose two distinct points, say b and c, lying in $\mathbb{C} \setminus D$. Then the Möbius transformation

$$\lambda(z) := \frac{z - b}{z - c}$$

is holomorphic on D and does not vanish at any point of D. Therefore, by Proposition 20.2 there exists $g_1 \in H(D)$ such that $g_1^2 = \lambda$, and we let $g_2 := -g_1$. Clearly, $g_2^2 = \lambda$.

Lemma 21.1. *We have*

(1) *each of the functions g_1, g_2 is 1-to-1;*
(2) $g_1(D) \cap g_2(D) = \emptyset$.

Proof. To obtain Part (1), suppose that for some $j \in \{1, 2\}$ and some $z_1, z_2 \in D$ one has $g_j(z_1) = g_j(z_2)$. Then $g_j^2(z_1) = g_j^2(z_2)$, that is, $\lambda(z_1) = \lambda(z_2)$, which implies $z_1 = z_2$ since λ is 1-to-1.

To establish Part (2), suppose that $g_1(D) \cap g_2(D) \neq \emptyset$. Hence, for some $z_1, z_2 \in D$ we have $g_1(z_1) = g_2(z_2)$. Therefore, $g_1^2(z_1) = g_2^2(z_2)$, i.e., $\lambda(z_1) = \lambda(z_2)$, which as before yields $z_1 = z_2$ thus leading to a contradiction. $\square$

Since $g_2 \not\equiv$ const, by Theorem 3.2 the set $g_2(D)$ is a domain. Fix $w_0 \in g_2(D)$ and find a positive number r such that $\Delta(w_0, r) \subset g_2(D)$. Then by Part (2) of Lemma 21.1 one has

$$|g_1(z) - w_0| \geq r \quad \forall z \in D.$$

Consider the function

$$g_0(z) := \frac{r}{2(g_1(z) - w_0)}.$$

Clearly, $g_0 \in H(D)$, and by Part (1) of of Lemma 21.1 g_0 is 1-to-1. Furthermore, $|g_0(z)| < 1$ for all $z \in D$.

Now, fix $z_0 \in D$ and consider the subset $\mathscr{F} \subset H(D)$ that consists of all functions $f \in H(D)$ satisfying the following conditions:

(1) f is 1-to-1;
(2) $|f(z)| < 1 \; \forall z \in D$;
(3) $|f'(z_0)| \geq |g_0'(z_0)|$.

Obviously, $\mathscr{F}$ contains g_0 and therefore is non-empty.

Lemma 21.2. *The set $\mathscr{F} \subset H(D)$ is compact.*

Proof. As $|f(z)| < 1$ for all $f \in \mathscr{F}$ and $z \in D$, by Theorem 18.2 the set $\mathscr{F}$ is pre-compact. Thus, in order to establish the lemma we only need to show that $\mathscr{F}$ is closed in $H(D)$. Let $\{f_n\}$ be a sequence in $\mathscr{F}$ convergent in $H(D)$ to some function F. We will now prove that $F \in \mathscr{F}$.

First, observe that

$$|F'(z_0)| \geq |g_0'(z_0)|. \tag{21.1}$$

Indeed, on the one hand, for all n we have

$$|f_n'(z_0)| \geq |g_0'(z_0)|.$$

On the other hand, by Example 18.1 the map

$$L : H(D) \to \mathbb{C}, \quad f \mapsto f'(z_0) \tag{21.2}$$

is a continuous linear functional on $H(D)$. Inequality (21.1) is an immediate consequence of these two facts.

Further, by Corollary 18.4 we see that either F is 1-to-1, or $F \equiv \text{const}$. We claim that the latter is not the case. Indeed, as g_0 is 1-to-1, Theorem 3.3 implies $g_0'(z_0) \neq 0$. Hence by inequality (21.1) we have $F'(z_0) \neq 0$, and therefore $F \not\equiv \text{const}$. This shows that F is 1-to-1.

Finally, as

$$|f_n(z)| < 1 \; \forall z \in D, \; \forall n,$$

we obtain

$$|F(z)| \leq 1 \; \forall z \in D.$$

Since $F \not\equiv \text{const}$, Theorem 17.3 yields

$$|F(z)| < 1 \; \forall z \in D.$$

Thus, $F \in \mathscr{F}$ as required. $\square$

Since $\mathscr{F}$ is compact by Lemma 21.2, from Proposition 18.1 it follows that there exists an element $f_0 \in \mathscr{F}$ such that

$$|L(f_0)| \geq |L(f)| \; \forall f \in \mathscr{F},$$

where L is the continuous linear functional on $H(D)$ introduced in (21.2). In other words, we have

$$|f_0'(z_0)| \geq |f'(z_0)| \ \forall f \in \mathscr{F}.$$

We will now prove that f_0 is a conformal map from D onto Δ.

As f_0 lies in $\mathscr{F}$, we know that f_0 is 1-to-1 and maps D *into* Δ. Therefore, it only remains to show that the map $f_0 : D \to \Delta$ is *onto*. In order to establish this, we need two lemmas. The first one concerns conformal transformations of Δ (see Proposition 5.5 and Corollary 18.1).

Lemma 21.3. *For $a \in \Delta$, consider the Möbius transformation of Δ*

$$\lambda_a(z) := \frac{z - a}{1 - \bar{a}z}. \tag{21.3}$$

Then

$$\lambda_a'(0) = 1 - |a|^2, \quad \lambda_a'(a) = \frac{1}{1 - |a|^2}.$$

Proof. Homework. $\square$

The other lemma that we require is:

Lemma 21.4. *One has $f_0(z_0) = 0$.*

Proof. Let $d := f_0(z_0)$ and assume that $d \neq 0$. Consider the composition of maps $F_1 := \lambda_d \circ f_0$. Clearly, F_1 is 1-to-1 and $|F_1(z)| < 1$ for all $z \in D$. Furthermore, by Lemma 21.3 we see

$$|F_1'(z_0)| = |\lambda_d'(d)| \cdot |f_0'(z_0)| = \frac{|f_0'(z_0)|}{1 - |d|^2} > |f_0'(z_0)| \geq |g_0'(z_0)|.$$

Hence, $F_1 \in \mathscr{F}$ and $|L(F_1)| > |L(f_0)|$, which contradicts our choice of f_0. Therefore $d = 0$. $\square$

We are now ready to show that f_0 maps D onto Δ thus finalising the proof of Theorem 8.3. Assume that $f_0 : D \to \Delta$ is not surjective and choose $p \in \Delta \setminus f_0(D)$. Notice that by Lemma 21.4, we have $p \neq 0$. Consider $F_2 := \lambda_p \circ f_0$. Clearly, F_2 is 1-to-1 and $|F_2(z)| < 1$ for all $z \in D$. Using Lemmas 21.3, 21.4 we compute

$$F_2'(z_0) = \lambda_p'(0)f_0'(z_0) = (1 - |p|^2)f_0'(z_0). \tag{21.4}$$

Next, as F_2 lies in $H(D)$ and does not vanish at any point of D, by Proposition 20.2 there exists $h \in H(D)$ such that $h^2 = F_2$. Observe that h is 1-to-1 (explain!) and $|h(z)| < 1$ for all $z \in D$. Since

$$F_2'(z_0) = (h^2)'(z_0) = 2h'(z_0)h(z_0),$$

from (21.4) we deduce

$$h'(z_0) = \frac{(1-|p|^2)f_0'(z_0)}{2h(z_0)},$$

hence

$$|h'(z_0)| = \frac{(1-|p|^2)|f_0'(z_0)|}{2\sqrt{|p|}}. \tag{21.5}$$

Further, let

$$h_0 := \lambda_{h(z_0)} \circ h.$$

Notice that h_0 is 1-to-1 and $|h_0(z)| < 1$ for all $z \in D$. Moreover, using Lemma 21.3 and formula (21.5) we compute

$$|h_0'(z_0)| = |\lambda_{h(z_0)}'(h(z_0))| \cdot |h'(z_0)| = \frac{|h'(z_0)|}{1-|h(z_0)|^2} =$$

$$\frac{(1-|p|^2)|f_0'(z_0)|}{2\sqrt{|p|}(1-|p|)} = \frac{(1+|p|)|f_0'(z_0)|}{2\sqrt{|p|}} > |f_0'(z_0)| \geq |g_0'(z_0)|,$$

where we utilised the inequality

$$\frac{(1+|p|)}{2\sqrt{|p|}} > 1$$

(prove it!). Thus, $h_0 \in \mathscr{F}$ and $|L(h_0)| > |L(f_0)|$, which contradicts our choice of f_0. Hence, the map $f_0 : D \to \Delta$ is onto as required.   □

Remark 21.1. Let us for the moment restrict our considerations to domains in $\mathbb{C}$. Such a domain D is called *holomorphically simply-connected* if for every $f \in H(D)$ and every closed path γ in D one has $\int_\gamma f dz = 0$. By Corollary 9.1, every simply-connected domain in $\mathbb{C}$ is holomorphically simply-connected. Now, notice that the proof of Theorem 8.3 given in this lecture works for all holomorphically simply-connected domains not equal to $\mathbb{C}$. Indeed, recall from Remark 20.1 that Proposition 20.2 can be proved by using Corollary 9.2. For a holomorphically simply-connected domain $D \subset \mathbb{C}$, $D \neq \mathbb{C}$, the result of Corollary 9.2 is valid (explain!), and therefore D is conformally equivalent to Δ. Thus, every holomorphically simply-connected domain either is all of $\mathbb{C}$ or is conformally equivalent to Δ and therefore is simply-connected (explain!). This shows that the class of simply-connected domains in $\mathbb{C}$ coincides with that of holomorphically simply-connected ones. We note that the above observation is utilised in the proof of the sufficiency implications of Theorems 8.1, 8.2.

We will now slightly extend Theorem 8.3 by including a statement on the uniqueness of a conformal map from a given simply-connected domain onto Δ.

Theorem 21.1. (The Riemann Mapping Theorem, Final Version) *Let $D \subset \overline{\mathbb{C}}$ be a simply-connected domain such that $\overline{\mathbb{C}} \setminus D$ contains at least two points. Then D is conformally equivalent to Δ. Moreover, for any point $z_0 \in D$, $z_0 \neq \infty$, a conformal map, say f, from D onto Δ can be chosen to satisfy the conditions*

(1) $f(z_0) = 0;$
(2) $f'(z_0) > 0,$

and these conditions determine a conformal equivalence between D and Δ uniquely.

Proof. By Theorem 8.3, there exists a conformal map, say f_0, from D onto Δ. Let $a := f_0(z_0)$ and consider the composition $F := \lambda_a \circ f_0$, where λ_a is the Möbius transformation of Δ introduced in (21.3). Clearly, we have $F(z_0) = 0$. By Theorem 4.1, the map F is $\mathbb{C}$-differentiable at z_0, and we set

$$\alpha := -\arg F'(z_0)$$

(note that $F'(z_0) \neq 0$). Then $f := e^{i\alpha} F$ satisfies Conditions (1) and (2) and maps D conformally onto Δ.

We will now show that a conformal map for which both (1) and (2) hold is unique. Indeed, let f_1, f_2 be any two conformal maps from D onto Δ. The composition $f_1 \circ f_2^{-1}$ is a conformal transformation of Δ and therefore, by Corollary 18.1, we see

$$f_1 \circ f_2^{-1} = e^{i\alpha} \lambda_a$$

for some $a \in \Delta$ and $\alpha \in \mathbb{R}$. Hence, we have

$$f_1 = e^{i\alpha} \lambda_a \circ f_2.$$

If $f_1(z_0) = f_2(z_0) = 0$, it is immediate that $a = 0$, which implies $f_1 = e^{i\alpha} f_2$. If, furthermore, $f_1'(z_0) > 0$ and $f_2'(z_0) > 0$, it follows that $e^{i\alpha} = 1$, thus $f_1 = f_2$ as required. $\square$

Theorem 21.1 yields a classification of simply-connected domains in $\overline{\mathbb{C}}$.

Corollary 21.1. *Every simply-connected domain in $\overline{\mathbb{C}}$ is conformally equivalent to one and only one of Δ, $\mathbb{C}$, $\overline{\mathbb{C}}$.*

Proof. Let $D \subset \overline{\mathbb{C}}$ be a simply-connected domain. If $\overline{\mathbb{C}} \setminus D$ contains at least two points, D is conformally equivalent to Δ by Theorem 21.1. Next, if $\overline{\mathbb{C}} \setminus D$ is a single point, we apply a Möbius transformation that maps this point to ∞. Such a transformation is a conformal equivalence between D and $\mathbb{C}$. Finally, if $\overline{\mathbb{C}} \setminus D = \emptyset$, then $D = \overline{\mathbb{C}}$.

It remains to show that the domains Δ, $\mathbb{C}$, $\overline{\mathbb{C}}$ are pairwise conformally non-equivalent. First of all, Δ and $\mathbb{C}$ are non-compact metric spaces whereas $\overline{\mathbb{C}}$ is a compact one (explain!). As any conformal equivalence is a homeomorphism, there exists no conformal map from $\overline{\mathbb{C}}$ onto either Δ or $\mathbb{C}$. Further, if f is a conformal map from $\mathbb{C}$ onto Δ, Theorem 4.1 yields that f is an entire function. Therefore, Theorem 13.1 implies $f \equiv \text{const}$, which is impossible as f is 1-to-1. $\square$

We close the course by determining all conformal transformations of the canonical simply-connected domains listed in Corollary 21.1. Note that the conformal transformations of Δ are given by Corollary 18.1, so we only need to describe those of $\mathbb{C}$ and $\overline{\mathbb{C}}$.

Proposition 21.1. *We have*

(1) *every conformal transformation of* $\mathbb{C}$ *has the form* $f(z) = az + b$ *with* $a, b \in \mathbb{C}$, $a \neq 0$;

(2) *every conformal transformation of* $\overline{\mathbb{C}}$ *is a Möbius transformation.*

Proof. To obtain Part (1), let f be a conformal transformation of $\mathbb{C}$. Then, by Theorem 4.1, the map f is an entire function, hence ∞ is an isolated singularity of f. We claim that ∞ is a pole. Indeed, if ∞ is a removable singularity, $|f(z)|$ is a bounded function on $\mathbb{C}$. Therefore, Theorem 13.1 implies $f \equiv$ const, which is impossible since f is 1-to-1. Next, if ∞ is an essential singularity, by Theorem 15.1 there exists a sequence $\{z_n\}$ converging to ∞ such that the sequence $\{w_n := f(z_n)\}$ converges, say, to 0. Clearly, $\{f^{-1}(w_n)\}$ converges to ∞. On the other hand, by Corollary 3.1, the inverse map f^{-1} is conformal on $\mathbb{C}$ and, in particular, is continuous at 0. This contradiction shows that ∞ is indeed a pole of f as claimed.

We will now make the following easy observation:

Lemma 21.5. *An entire function having a pole at* ∞ *is a polynomial in z of positive degree.*

Proof. Let g be an entire function having a pole at ∞. Write g as the sum of a power series centred, say, at 0

$$g(z) = \sum_{n=0}^{\infty} c_n z^n.$$

By Part (2) of Theorem 14.3, the series is finite and non-constant, i.e., g is indeed a polynomial in z of positive degree. $\quad\square$

By Lemma 21.5, the function f is a polynomial in z of degree $K \geq 1$. Then Theorem 1.1 yields that f has exactly K zeroes in $\mathbb{C}$ where each zero is counted with its order. Since f is 1-to-1, all the zeroes must coincide, i.e.,

$$f(z) = a_K (z - w_0)^K$$

for some $a_K, w_0 \in \mathbb{C}$ with $a_K \neq 0$. But $f'(w_0) = 0$ if $K \geq 2$, which contradicts Theorem 3.3 as f is 1-to-1. Hence $K = 1$, and Part (1) is established.

To obtain Part (2), let λ be a Möbius transformation such that $\lambda(f(\infty)) = \infty$. Then $\lambda \circ f|_{\mathbb{C}}$ is a conformal transformation of $\mathbb{C}$ and therefore is an affine transformation by Part (1). Thus, f is a Möbius transformation. $\quad\square$

We will now give algebraic descriptions of the groups of conformal transformations (with respect to composition) of the canonical simply-connected domains listed in Corollary 21.1. Recall from Corollary 18.3 that the group of conformal transformations of Δ is isomorphic to the group $\text{PSL}_2(\mathbb{R})$ defined in Corollary 5.3. To complement this result, we state the following consequence of Proposition 21.1:

Corollary 21.2. *We have:*

(1) *the group of conformal transformations of* $\mathbb{C}$ *is isomorphic to the semi-direct product* $\mathbb{C}^* \ltimes \mathbb{C}$, *where* $\mathbb{C}^*$ *is the multiplicative group of non-zero complex numbers acting on the normal subgroup* $\mathbb{C}$ *by multiplication;*

(2) *the group of conformal transformations of* $\overline{\mathbb{C}}$ *is isomorphic to* $\mathrm{PGL}_2(\mathbb{C})$, *which is the group defined in Corollary* 5.1.

Proof. Homework. □

Exercises

21.1. Let $D \subset \overline{\mathbb{C}}$ be a simply-connected domain such that $\overline{\mathbb{C}} \setminus D$ contains at least two points. Prove that for any point $a \in \Delta$, a number $0 \le \alpha < 2\pi$, and any point $z_0 \in D$, $z_0 \ne \infty$, a conformal map, say f, from D onto Δ can be chosen to satisfy the conditions

(1) $f(z_0) = a$;
(2) $\arg f'(z_0) = \alpha$,

and these conditions determine a conformal equivalence between D and Δ uniquely.

21.2. Let f be a conformal transformation of an annulus $\Delta(0, r_1, r_2)$, where $0 < r_1 < r_2 < \infty$. Assume that f extends to a homeomorphism of $\overline{\Delta(0, r_1, r_2)}$. Prove that either $f(z) = cz$, or $f(z) = c/z$ for some $c \in \mathbb{C} \setminus \{0\}$. What values of c are allowed in each of the two cases? (Hint: in the spirit of Exercise 2.8, use symmetry with respect to circles to produce a holomorphic extension of f beyond $\Delta(0, r_1, r_2)$.)

21.3. Show that two annuli $\Delta(0, r_1, r_2)$ and $\Delta(0, \tilde{r}_1, \tilde{r}_2)$, where $0 < r_1 < r_2 < \infty$, $0 < \tilde{r}_1 < \tilde{r}_2 < \infty$, are conformally equivalent by means of a map

$$f : \Delta(0, r_1, r_2) \to \Delta(0, \tilde{r}_1, \tilde{r}_2)$$

that extends to a homeomorphism from $\overline{\Delta(0, r_1, r_2)}$ onto $\overline{\Delta(0, \tilde{r}_1, \tilde{r}_2)}$, if and only if $r_2/r_1 = \tilde{r}_2/\tilde{r}_1$.

21.4. Find all conformal transformations of the complement $\Delta(0, r, \infty)$ to the closed disk $\overline{\Delta_r}$, with $0 < r < \infty$, that extend to homeomorphisms of the closure $\overline{\Delta(0, r, \infty)}$ in $\mathbb{C}$.

21.5. Find all conformal transformations of the punctured complex plane $\mathbb{C} \setminus \{0\}$. (Hint: study the isolated singularity of a conformal transformation of $\mathbb{C} \setminus \{0\}$ at 0.)

21.6. Find all conformal transformations of the punctured disk $\Delta \setminus \{0\}$.

21.7. Give an example of a bounded Jordan domain whose boundary has more than two components that possesses conformal transformations not equal to the identity. Can you find an example with a family of pairwise distinct conformal transformations depending on a continuous parameter?

21.8. For any $n = 2, 3, \ldots$, give an example of a bounded Jordan domain whose boundary has more than two components and whose group of conformal transformation contains a subgroup isomorphic to the group of roots of 1 of order n.

21.9. Let $D \subset \mathbb{C}$ be a domain and $a_1, \ldots, a_n \in D$. Assume that $f \in H(D \setminus \{a_1, \ldots, a_n\})$ and that each a_j is a pole of f. Let g_j be the principal part of the Laurent series expansion of f with centre a_j, $j = 1, \ldots, n$. Prove that the function

$$f - \sum_{j=1}^{n} g_j,$$

which lies in $H(D \setminus \{a_1, \ldots, a_n\})$, extends to a function holomorphic on D.

21.10. Let $a_1, \ldots, a_n$ be points in $\mathbb{C}$ and $f \in H(\mathbb{C} \setminus \{a_1, \ldots, a_n\})$. Suppose that each of $a_1, \ldots, a_n$ is a pole of f. Assume that ∞ is a pole of f as well. Prove that f is a rational function, that is, there exist two polynomials $P(z)$ and $Q(z)$ in z such that

$$f(z) = \frac{P(z)}{Q(z)}.$$

(Hint: use Exercise 21.9.)

21.11. Let $D \subset \mathbb{C}$ be a domain and $a \in D$. Assume that $f \in H(D \setminus \{a\})$ and let g be the principal part of the Laurent series expansion of f with centre a. Consider the function $h := f - g$ on a punctured disk centred at a and contained in D. Does h extend to a function holomorphic on D? Prove your conclusion.

Index

Printed in the United States
By Bookmasters